Kalman Filter - Engineering Applications

Edited by Yuri V. Kim

Published in London, United Kingdom

Kalman Filter - Engineering Applications
http://dx.doi.org/10.5772/intechopen.100722
Edited by Yuri V. Kim

Contributors
Yuri V. Kim, Xiaoxiao Sun, Joel Parker, Yifan Zhang, Bonnie J. LaFleur, Rajamani Doraiswami, Lahouari Cheded, Alexandre Schalch Mendes, Janito Vaqueiro Ferreira, Pablo Siqueira Meirelles, Eduardo Rodrigues de Lima, Hee-Seung Kim, Adam Bienkowski, David Sidoti, Krishna R. Pattipati, Yaakov Bar-Shalom, David L. Kleinman, Lingyi Zhang

First published in London, United Kingdom, 2023 by IntechOpen
IntechOpen is the global imprint of INTECHOPEN LIMITED, registered in England and Wales, registration number: 11086078, 5 Princes Gate Court, London, SW7 2QJ, United Kingdom

British Library Cataloguing-in-Publication Data
A catalogue record for this book is available from the British Library

Additional hard and PDF copies can be obtained from orders@intechopen.com

Kalman Filter - Engineering Applications
Edited by Yuri V. Kim
p. cm.
Print ISBN 978-1-80356-575-0
Online ISBN 978-1-80356-576-7
eBook (PDF) ISBN 978-1-80356-577-4

Meet the editor

Yuri V. Kim graduated from the Moscow Aviation Institute as an electro-mechanical engineer specializing in aerospace GN&C and avionics. He obtained his Ph.D. and Doctor of Technical Science degrees in aerospace GN&C. After graduation, he worked in the aerospace industry and universities of the USSR as an engineer, research scientist, manager and professor. After the collapse of the USSR, he worked for the Israel Aerospace Industries Corporation (AIA) and then for the Canadian Space Agency as an engineer and scientist, lecturing and training Canadian students and new engineers and scientists. His main areas of interest are state estimation and multisensory navigation systems. For the last several years he has been working on new methods for satellite navigation and control, and functional qualification tests for satellite control systems. He has many scientific publications and inventions that have been implemented in flight systems in the USSR, Israel and Canada.

Contents

Preface

The purpose of this book is to present some uses of the Kalman filter (KF) in engineering activities that can produce a robust and technically acceptable result while keeping as close as possible to the optimal (most accurate) solution. KF sub-optimization is often required, due to the realities of implementation and real-life operational conditions. The book brings together the experiences of specialists from different engineering areas using the KF in their practice.

Chapter 1 discusses the use of the sub-optimal KF (filter with bounded growth of memory, FBGM) for the analytical design of a navigation accelerometer with an electric spring. Two approaches are compared: empirical and analytic (using sub-optimal KF). The comparison demonstrates that conscious use of the analytical design approach can enable even the experienced beginner engineer to achieve an appropriate and close to an optimal result.

Chapter 2 examines the application of the extended Kalman filter (EKF) to predict the movement of the upper top center of a guyed tower of a power line, based on its dynamic model and considering the indirect measurement of the force in the stay cables through accelerometers. The proposed methodology is validated through experiments carried out in the mockup tower of a dynamic testing laboratory.

Chapter 3 discusses an intelligent integrated KF approach, which combines the model based approach (MBA) and the model-free approach (MFA) for the diagnostic of a pipeline failure. The proposed scheme ensures a fast and accurate diagnosis and was successfully used on a two-tank model in a laboratory.

Chapter 4 describes a computationally efficient approach to using smoothing spline models for filtering and analyzing complex signals from noisy data. Estimation algorithms based on the Kalman filter are implemented within an innovative smoothing spline analysis of variance (SS-ANOVA) model. This approach can be used in different engineering areas, where experimental data are interpolated by a spline for further development of a system model.

Chapter 5 presents a sequential mini-batch noise estimator to estimate the covariance of measured signal noise. This problem is typical of using the KF in various applications. The proposed method, which allows for estimation that is both sufficient and computationally economical, is demonstrated by several engineering problems.

I am thankful to all the authors for their contributions and to IntechOpen for its decision to publish this book and for help with my editorial work. I would also like to express my gratitude to the Canadian Space Agency for its friendly attitude and the moral support that has encouraged me to present this book to the international engineering and scientific community.

Yuri V. Kim
David Florida Laboratory,
Canadian Space Agency,
Ottawa, Canada

Chapter 1

Review of Kalman Filter Developments in Analytical Engineering Design

Yuri V. Kim

Abstract

This chapter discusses using the Kalman Filter (KF) for the analytical design in such engineering applications, as a closed loop control system, which often can be considered as time-invariant and linear (LTI). The chapter discuses designing a navigation accelerometer with the electric spring. This is a typical example of the closed loop, negative feedback control. Two approaches are used: A-conventional, empirical and B-analytical with KF. The consideration of both of them opens a comprehensive understanding of the system dynamics and its potentials. The discussion is based on the suboptimal form of the KF-Filter with Bounded Growth of Memory (FBGM), proposed by the author.

Keywords: dynamic system, Kalman Filter, covariance, standard deviation, Riccati equation, stochastic process, white noise, optimal estimation and control, analytical design, accelerometer

1. Introduction

Appearance the Kalman Filter in 60-th [1, 2] led in the control system engineering to a trend that continues even today; implementing this powerful a scientific tool directly into a real system. This tendency is based on the "doctrine" of considering the KF, as a "magic box", which is capable to achieve a superior performance sometimes even with attempt for estimation not observable system state variables. However, directly implementing KF in the "real word system", a developer may often face many unexpected issues, resulted in the negative conclusion about its practical applicability.

Both situations are extreme. A conscious application of KF with premature engineer analysis and sub-optimization, following by sacrificing the theoretically achievable maximal accuracy for the expense of the filter robustness and the simplicity can lead to the applicable results, similar to the conventional solution. More fine tuning the control gains in this case allows for getting better performance. This approach can be presented as a "rule of thumb" – meeting performance requirements, rather than achieving theoretical excellence was presented in many works of the author: [3–11]. In particular, some simple KF sub-optimization form – Kalman-Bucy filter (KBF), with bounded grows of memory (FBGM) was proposed in [8, 9].

FBGM guarantees accurate system performance, maintaining the continuity of the conventional solutions. Specifically, following by this approach is an appropriate in the case of the Linear Time Invariant (LTI) systems. Often, for the LTI system fast estimation and insertion of the initial conditions are not required and the KF can be used in its stationary (LTI) form with constant coefficients that can be found solving KF Ricatti equation in the steady state, when it degenerates in the algebraic form. If the main single criterion of system design quality can be formulated as the minimum of system error standard deviations (STD), then the KF can be used as the filter and the feedback closed loop controller simultaneously. Namely, this case is discussed in the example, presented in the chapter below.

For many similar examples the empirical engineer solutions are well-known for years and have become conventional, demonstrating a good performance and robustness in a wide range of possible operational conditions. However, using the KF and the analytical design is always useful to check available potentials and is specifically necessary, when there is a lack of experience in similar design. In the real life many factors usually play essential role for the system (devise) design. Indeed, some of them can hardly be mathematically formalized and put restrictions for a "freedom" of the analytical solution. However, often a single criterion for the considered system can be taking as a dominant –minimum steady state errors covariance matrix, which is the optimization criterion the optimal KF. That is why if at the first stage of system design developers are advised to use this criterion and the KF as a helpful tool for the analytical design.

The basic information about the KF can be found in [1, 2, 12–16], about the Analytical Design (AD) – in [9, 17–20]. This AD approach is well compatible with System Engineering Model Based Design, providing with Matlab/Simulink Tools by the MathWorks American Company [21, 22].

It must be said, in addition, that due to essential progress in computerization, achieved up to date and implementing of the KF in a broad spectrum of new applications such as system identification, failure detection, image and data processing applied research in physics and mathematics (some of them are presented in this book), direct implementing of the KF algorithm and its further modifications has become common and justifiable remedy for many developers. MathWorks Company provides powerful and universal Simulink KF algorithms [23] that can be used on different stages of the research and implementation.

2. Kalman filter

As it has been mentioned in the introduction, the KF was proposed by Rudolf Kalman in 1960 [1]. From the physical point of view it was just a filter that can separate deterministic- stochastic signal from the measurement stochastic noise. However, the mathematical form of the presentation as a vector-matrix space state differential equation (many available inputs and outputs) and provided optimal criterion –minimum of the estimation covariance matrix errors made this filter be used as a powerful tool for a scientific research and engineering development in many areas. A big advantage of the filter is that it was presented as a discrete algorithm allowing for the recursive computation. It allows estimating variable at the current step, using its predicted estimation, based on the system model prediction (without the measurements). This estimation is taken with the control gain (weight coefficient) proportional to the estimation errors covariance matrix and in inverse proportion to the matrix of the measured noise. This coefficient is multiplied by the difference between the measured vector and its estimation. The discrete covariance matrix was provided allowing for the recursive solution of the non-linear matrix Riccati equation. Thus these equations of the filter, presented by Kalman for the discrete stochastic process [1], opened direct way to compute estimates and implement the filter in the form of the computation algorithm. Further many authors (for example, [12, 14]) presented these the discrete KF equations with detail explanation of the filter and the computation process. Pseudo-code of the filter provided by MathWorks (Simulink) for a broad pool of international users is presented in [23]. Following [12], generic pseudo-code of KF can be also presented as below.

Let us given the stochastic model of a linear dynamic system

$$\dot{\mathbf{x}}_{i+1} = \mathbf{\Phi}_i \mathbf{x}_i + \mathbf{G}_i \mathbf{w}_i, \ \ i = 0,1,\dots,N-1 \tag{1}$$

where $\mathbf{x}_i$ is system state vector, i is computation step number, $\mathbf{\Phi}_i$ is system pulse transfer function, $\mathbf{G}_i$ is system input impact function, $\mathbf{w}_i$ is system input impact stochastic process

Every step system state vector is measured as follows

$$\mathbf{z}_i = \mathbf{H}_i \mathbf{x}_i + \mathbf{v}_i \tag{2}$$

where $\mathbf{z}_i$ is measurement vector, $\mathbf{H_i}$ is measurement matrix, $\mathbf{v}_i$ is stochastic measurement error. The processes $\mathbf{w}_i$ and $\mathbf{v}_i$ are centered mutually non-correlated white Gaussian noises, having the covariance matrixes $\mathbf{Q_i}$ and $\mathbf{R_i}$ respectively. The KF at each step will provide the optimal estimate $\hat{\mathbf{x}}_\mathbf{i}$ of the system state vector $\mathbf{x_i}$ with the minimum of the estimation errors $\tilde{\mathbf{x}}_\mathbf{i} = \hat{\mathbf{x}}_\mathbf{i} - \mathbf{x_i}$ covariance matrix $\mathbf{P_i} - \mathbf{E}\left[\tilde{\mathbf{x}}_\mathbf{i} \cdot \tilde{\mathbf{x}}_\mathbf{i}^\mathbf{T}\right]$

Then for the system (1), (2) KF pseudo-code is as follows

It has been mentioned in the introduction that in many occurrences direct implementation of the algorithm above may lead to the optimal filter instability and, at least, to essential requirements for the computational resources. However, sub-optimal form of the filter may solve the problem within required accuracy, but with warranted stability, robustness and very modest requirements for the computation (often even analogue devices). One of such sub-optimization, developed by the author, is presented below (**Figure 1**).

3. Kalman-Busy filter sub-optimization

Let us consider the analog Kalman filter form that was presented by Kalman and Busy in [2]. Given is linear, fully observable and controllable stochastic system

$$\begin{aligned} \dot{\mathbf{x}} &= \mathbf{F}\mathbf{x} + \mathbf{G}\mathbf{w}, \\ \mathbf{z} &= \mathbf{H}\mathbf{x} + \mathbf{v}, \end{aligned} \tag{3}$$

where: $\mathbf{x}$ is an n – vector of the system state, $\mathbf{F}$ is $n \times n$ system dynamics matrix, $\mathbf{w}$ is an n – vector of external disturbances, $\mathbf{G}$ is an $(n \times n)$ disturbances matrix, $\mathbf{z}$ is a $m \leq n$ is vector of measurements, $\mathbf{v}$ is a m is vector of measurement noise, $\mathbf{H}$ is a $(m \times n)$ measurements matrix.

Input: $\mathbf{\Phi}_i, \mathbf{H}_i, \mathbf{Q}_i, \mathbf{R}_i, \mathbf{z}_i$

Output : $\hat{\mathbf{x}}_{i+1}, \mathbf{P}_{i+1}$

Step 1 – Set intial model parameters

$\mathbf{\Phi}_0, \mathbf{H}_0, \mathbf{Q}_0, \mathbf{R}_0$

Step 2 – Set initial conditions

$\hat{\mathbf{x}}_0, \mathbf{P}_0$

and initialise the estimation circle (i=0,1,..N)

Step 3-Compute pedicted (a priori) estimate of the state vector $\bar{\mathbf{x}}_{i+1}$ and covariance matrix $\mathbf{M}_{i+1}$

$\bar{\mathbf{x}}_{i+1} = \mathbf{\Phi}_i \hat{\mathbf{x}}_i$

$\mathbf{M}_{i+1} = \mathbf{\Phi}_i \mathbf{P}_i \mathbf{\Phi}_i^T + \mathbf{G}_i \mathbf{Q}_i \mathbf{G}_i^T$

Step 4 – Compute updated (a posteriori) covariance matrix $\mathbf{P}_{i+1}$ and KF weight coefficient $\mathbf{K}_{i+1}$

$\mathbf{P}_{i+1} = \mathbf{M}_{i+1} - \mathbf{M}_{i+1}\mathbf{H}_{i+1}^T(\mathbf{H}_{i+1}\mathbf{M}_{i+1}\mathbf{H}_{i+1}^T + \mathbf{R}_i)^{-1}\mathbf{H}_{i+1}\mathbf{M}_{i+1}$

$\mathbf{K}_{i+1} = \mathbf{P}_{i+1}\mathbf{H}_{i+1}^T\mathbf{R}_{i+1}^{-1}$

Step 5 - Compute updated (a posteriori) estimate of the state vector $\hat{\mathbf{x}}_{i+1}$

$\hat{\mathbf{x}}_{i+1} = \bar{\mathbf{x}}_{i+1} + \mathbf{K}_{i+1}(\mathbf{z}_{i+1} - \mathbf{H}_{i+1}\bar{\mathbf{x}}_{i+1})$

Step 6 – Repeat the circle with i:=i+1

if i=N, then end

Figure 1.
Discrete Kalman Filter algorithm pseudo-code.

Let us the following information about (3) is given: **F**,**G**,**H** are known matrices of time, these in the stationary case are matrix constants and

$$\begin{aligned}
&\mathbf{E}[\mathbf{x}(t_0)] = \mathbf{0},\ \mathbf{E}[\mathbf{w}(t)] = \mathbf{0},\ \mathbf{E}[\mathbf{v}(t)] = \mathbf{0},\\
&\mathbf{E}\left[\mathbf{x}(t_0)\mathbf{x}^{\mathrm{T}}(\mathrm{t}_0)\right] = \mathbf{P}_0,\\
&\mathbf{E}\left[\mathbf{w}(\mathrm{t})\mathbf{v}^{\mathrm{T}}(\tau)\right] = \mathbf{E}\left[\mathbf{v}(\mathrm{t})\mathbf{w}^{\mathrm{T}}(\tau)\right] = \mathbf{E}\left[\mathbf{w}(\mathrm{t})\mathbf{x}^{\mathrm{T}}(\mathrm{t}_0)\right] = \mathbf{0},\\
&\mathbf{E}\left[\mathbf{w}(t)\mathbf{w}^{\mathrm{T}}(\tau)\right] = \mathbf{Q}(t)\boldsymbol{\delta}(t-\tau),\\
&\mathbf{E}\left[\mathbf{v}(\mathrm{t})\mathbf{v}^{\mathrm{T}}(\tau)\right] = \mathbf{R}(t)\boldsymbol{\delta}(t-\tau),
\end{aligned} \tag{4}$$

where: $\mathbf{P}_0$ is the initial state covariance matrix, $\mathbf{R}(t)$ is the covariance matrix of measurement noise, $\mathbf{Q}(t)$ is the covariance matrix of disturbance noise, $\boldsymbol{\delta}(t-\tau)$ is the Dirac delta function. Hence, $\mathbf{w}(t)$ and $\mathbf{v}(t)$ are Gauss white noise processes. Usually, matrixes **Q** and **R** are diagonal and in the stationary case, these matrices are constants. They have meaning of the band bounded spectral densities of practically correlated, "colour" noises $\mathbf{w}(t)$ and $\mathbf{v}(t)$, correspondingly, that within their band bounds can be approximately taken as white

$$\begin{aligned} \mathbf{Q} &= \mathbf{S_w} = diag\left[\sigma_{wi}^2 \Delta t_w\right], {}_{i=1,2,..n} \\ \mathbf{R} &= \mathbf{S_v} = diag\left[\sigma_{vj}^2 \Delta t_v\right], {}_{j=1,2,..m} \end{aligned} \tag{5}$$

where σ_{wi},Δt_w and σ_{vj},Δt_v are the standard deviations (STD) and the sampling times (Δt) of these stochastic processes **w** and **v** correspondently. In KBF theory are idealistically presents like white noises with zero Δt and infinite covariance functions.

This system state (state vector **x**) at any time instant t can be estimated (found the optimal estimate $\hat{\mathbf{x}}$ for) by providing the minimum of system state estimation error ($\tilde{\mathbf{x}} - \hat{\mathbf{x}} - \mathbf{x}$) for the covariance matrix $\mathbf{P} = E\left[\tilde{\mathbf{x}}(\mathbf{t}) \cdot \tilde{\mathbf{x}}^T(\mathbf{t})\right]$ diagonal $J = diag\mathbf{P}$. In other words, KBF filter merit criterion is as follows

$$J_{\min} = \min\left(diag\mathbf{P}\right) \tag{6}$$

This criterion is provided when the KBF is used for estimation of the state vector of the system (3). The KBF is usually presented by the following matrix equations

$$\begin{aligned} &\dot{\hat{\mathbf{x}}} = \mathbf{F}\hat{\mathbf{x}} + \mathbf{K}(\mathbf{z} - \mathbf{H}\hat{\mathbf{x}}),\ \hat{\mathbf{x}}_0, \\ &\mathbf{K} = \mathbf{P}\mathbf{H}^T\mathbf{R}^{-1}, \\ &\dot{\mathbf{P}} = \mathbf{F}\mathbf{P} + \mathbf{P}\mathbf{F}^T - \mathbf{P}\mathbf{H}^T\mathbf{R}^{-1}\mathbf{H}\mathbf{P} + \mathbf{G}\mathbf{Q}\mathbf{G}^T, \mathbf{P}_0 \end{aligned} \tag{7}$$

where: $\mathbf{K} = \mathbf{K}(t)$ is the KBF weigh matrix gain, $\mathbf{P} = \mathbf{P}(t)$ is the KBF estimate errors covariance matrix that can be found from the solution of the third matrix equation in (7) (Riccati type equation). Eq. (7) presumes that measurement vector **z** is continuously available, and then (7) presents KBF in the, so called, "filtering mode". However, if at some time instance t_p the measurement process is ended or temporarily interrupted and the vector measurements **z** since then is not available, then the KBF can be transitioned in the "prediction" mode. This mode is obtained from the filtering mode by setting in (7) KBF matrix gain to zero $\mathbf{K}(t) \equiv 0$, when $t \geq t_p$

$$\begin{aligned} &\dot{\hat{\mathbf{x}}} = \mathbf{F}\hat{\mathbf{x}},\ \hat{\mathbf{x}}_0 = \hat{\mathbf{x}}_\mathbf{p}, \\ &\mathbf{K} \equiv \mathbf{0}, \\ &\dot{\mathbf{P}} = \mathbf{F}\mathbf{P} + \mathbf{P}\mathbf{F}^T, \mathbf{P}_0 = \mathbf{P}_\mathbf{p} \end{aligned} \tag{8}$$

To implement KBF in practice, especially in real time on-board computer (OBC) with limited computational capabilities, it is always useful to sub-optimize the filter, sacrificing potentially achievable maximal accuracy, for simplicity and robustness (*quid pro quo*), being satisfied by some tolerate level of (6) instead of exact minimum at any considered time instance. As author presented in his past publications [3, 4, 8, 9] KBF in the time domain can be equivalently decomposed into two filters, working in parallel; non-stationary KBF with time variant matrix coefficient $\tilde{\mathbf{K}}(t)$ and stationary KBF with time invariant (constant) coefficient $\mathbf{K}^*$. Both can be determined pre-calculated in advance, before using KBF in real time. As it was showed in author's works these parallel filters can be approximately represented as the consecutive – Filter With

Bounded Grows of Memory (KFBGM) . The results of application of optimal KBF and sub-optimal filter (FBGM), represented below are close.

Both coefficients $\tilde{\mathbf{K}}$ and $\mathbf{K}^*$ can be determined by the solving of modified KBF Riccati matrix equations for covariance matrixes $\mathbf{P}^{\sim}(t)$ and $\mathbf{P}^*$. This KBF modification is as follows

$$\begin{cases} \dot{\hat{\mathbf{x}}}^{\sim} = \mathbf{F}^*\hat{\mathbf{x}}^{\sim} + \tilde{\mathbf{K}}(\mathbf{z} - \mathbf{H}\hat{\mathbf{x}}^{\sim}), \ \hat{\mathbf{x}}^{\sim}{}_0, \\ \tilde{\mathbf{K}} = \tilde{\mathbf{P}}\mathbf{H}^{\mathrm{T}}\mathbf{R}^{-1}, \\ \dot{\tilde{\mathbf{P}}} = \mathbf{F}^*\tilde{\mathbf{P}} + \tilde{\mathbf{P}}\mathbf{F}^* - \tilde{\mathbf{P}}\mathbf{H}^{\mathrm{T}}\mathbf{R}^{-1}\mathbf{H}\tilde{\mathbf{P}}, \tilde{\mathbf{P}}_0 = \mathbf{P}_0 - \mathbf{P}^*, \\ \mathbf{F}^* = \mathbf{F} - \mathbf{K}^*\mathbf{H}, \\ \dot{\hat{\mathbf{x}}}^* = \mathbf{F}\hat{\mathbf{x}}^* + \mathbf{K}^*(\mathbf{z} - \mathbf{H}\hat{\mathbf{x}}^*), \hat{\mathbf{x}}_0^*, \\ \mathbf{K}^* = \mathbf{P}^*\mathbf{H}^{\mathrm{T}}\mathbf{R}^{-1}, \\ \mathbf{F}\mathbf{P}^* + \mathbf{P}^*\mathbf{F}^{\mathrm{T}} - \mathbf{P}^*\mathbf{H}^{\mathrm{T}}\mathbf{R}^{-1}\mathbf{H}\mathbf{P}^* + \mathbf{G}\mathbf{Q}\mathbf{G}^{\mathrm{T}} = \mathbf{0} \\ \hat{\mathbf{x}} = \hat{\mathbf{x}}^{\sim} + \hat{\mathbf{x}}^*, \hat{\mathbf{x}}^{\sim}(t \to \infty) \to \mathbf{0}, \hat{\mathbf{x}}(t \to \infty) \to \mathbf{x}^* \\ \mathbf{P} = \tilde{\mathbf{P}} + \mathbf{P}^*, \tilde{\mathbf{P}}(t \to \infty) \to \mathbf{0}, P(t \to \infty) \to \mathbf{P}^* \end{cases} \tag{9}$$

Original Riccati equation for the matrix $\mathbf{P}$ is split here into two equivalent equations: the differential equation for the transfer state $\tilde{\mathbf{P}}$ (3-d of (9)) and the algebraic one (7-th of (9)) for the steady state $\mathbf{P}^*$. For many practical applications takes place the following inequality

$$diag\mathbf{P}_0 >> diag\mathbf{P}^* \tag{10}$$

In this case, as it was shown in [3], two parallel KBF filters (9) can be approximately substituted by the single suboptimal filter, working consequently in time in two modes: at the beginning in "initial filtering" (IF) mode for the "quasi- deterministic" system (assuming that in (3) $\mathbf{w} = \mathbf{0}$ and $\mathbf{Q} = \mathbf{0}$) with time-variant (variable) gain $\tilde{K}(t)$ gate and after is automatically switched to "steady filtering" (SF) for the substantially stochastic system model with time-invariant (constant) gain $\mathbf{K}^*$ (assuming that after some IF period t_*, $t \geq t_*$, the transfer process practically entirely completed and SF can start). Unlike the KBF, that is a filter with unbounded grooving memory (assuming continuous solving of KBF Riccati equation to determine KBF gate taking into account for it all process prehistory), this suboptimal modification was named the "Filter with Bounded Growing Memory" (FBGM). The filter equations are presented by (11) and (12) below

$$\dot{\hat{\mathbf{x}}} = \mathbf{F}\hat{\mathbf{x}} + \mathbf{K}(\mathbf{z} - \mathbf{H}\hat{\mathbf{x}}),$$
$$\mathbf{K} = \begin{cases} \tilde{\mathbf{K}}, & t_0 \leq t \leq t_*, \\ \mathbf{K}^*, & t > t_*, \\ \mathbf{0}, & if\ z \equiv 0 \end{cases} \tag{11}$$

where: t_* is the time, required for unbiased estimation of all n components of vector $\mathbf{x}$, when covariance matrix $\tilde{\mathbf{P}}$ decaying to a small matrix $\tilde{\mathbf{P}}(t_*) \approx \mathbf{0}$. The gains $\tilde{\mathbf{K}}$ and $\mathbf{K}^*$ are found with the following formulas

$$\begin{cases} \tilde{\mathbf{K}} = \tilde{\mathbf{P}}\mathbf{H}^{\mathrm{T}}\mathbf{R}^{-1}, \\ \dot{\tilde{\mathbf{P}}} = \mathbf{F}\tilde{\mathbf{P}} + \tilde{\mathbf{P}}\mathbf{F}^{\mathrm{T}} - \tilde{\mathbf{P}}\mathbf{H}^{\mathrm{T}}\mathbf{R}^{-1}\mathbf{H}\tilde{\mathbf{P}},\ \tilde{\mathbf{P}}_0 = \mathbf{P}_0, \\ \mathbf{K}^* = \mathbf{P}^*\mathbf{H}^{\mathrm{T}}\mathbf{R}^{-1}, \\ \mathbf{F}\mathbf{P}^* + \mathbf{P}^*\mathbf{F}^{\mathrm{T}} - \mathbf{P}^*\mathbf{H}^{\mathrm{T}}\mathbf{R}^{-1}\mathbf{H}\mathbf{P}^* + \mathbf{G}\mathbf{Q}\mathbf{G}^{\mathrm{T}} = \mathbf{0}. \end{cases} \tag{12}$$

In other words, $\tilde{\mathbf{K}}$ is computed for system (1), considered as a "quasi- deterministic"

($\mathbf{w} = \mathbf{0}, \mathbf{Q} = \mathbf{0}$, only transfer process, caused by $\mathbf{x}_0$ takes place), and $\mathbf{K}^*$ - considering (1) as a "substantially- stochastic" system, where only steady state motion, caused by the random disturbance $\mathbf{w}$ takes place, however the transfer process has been approximately decayed.

Three modes can be considered for the FBGM: IF- when $\mathbf{K} = \tilde{\mathbf{K}}$, SF- when $\mathbf{K} = \mathbf{K}^*$, and the "prediction mode" P-when measurement vector $\mathbf{z}$ is not available and $\mathbf{K} = \mathbf{0}$.

This filter can be easily periodically restarted after any interruption (outage) in measuring process. In the work [4] was introduced the "*observability index*" χ_i (power of signal to power of noise ratio) for certain $i - th$ component of the signal measured from the estimated system (1), considered as a quasi-deterministic. At the transition time t_* when KBF is switched from the IF mode to the SF mode all the observability indexes χ_i become bigger than one ($\chi_i >> 1,\ i = 1,2,\dots n$). This index is helpful to use it for FBGM analysis in the IF mode.

For filter analysis in the SF mode also can be introduced special "*filterability index*" $\xi_i = \frac{q_j}{r_i}, j = 1,2,..m; i = 1,2,..n$ – ratio of spectral density of signal exciting noise (Q) to spectral density of measured error noise ($\mathbf{R}$) for each pair of these noise components (w_j and v_i correspondingly). That index is similar to the observability index χ_i and is helpful for the FBGM analysis in SF mode.

Further developing the idea of FBGM (11), (12) is a new, simple KBF suboptimal modification FBGM-Initial/Steady (FBGM-I/S) can be considered. This idea is very simple, to change the time variable FBGM coefficient $\tilde{\mathbf{K}}(t)$, acting only during the transient period of the estimation process, by a permanent coefficient $\mathbf{K}^*_{\mathbf{IF}}$ that would accelerate it (comparatively to if the steady state coefficient $\mathbf{K}^*$ had been applied without $\tilde{\mathbf{K}}$ from the very beginning).

The assumption is that during this time the potentially available with KBF accuracy can be sacrificed for making this period shorter.

This modification is as follows

$$\dot{\hat{\mathbf{x}}} = \mathbf{F}\hat{\mathbf{x}} + \mathbf{K}(\mathbf{z} - \mathbf{H}\hat{\mathbf{x}}),$$

$$\mathbf{K} = \begin{cases} \mathbf{K}^*_{\mathbf{IF}}, & t_0 > t > t_*, \\ \mathbf{K}^*_{\mathbf{SF}}, & t > t_*, \\ \mathbf{0}, & \textit{if}\ \mathbf{z} \equiv 0, \end{cases} \tag{13}$$

where $\mathbf{K}^*_{\mathbf{IF}}$ and $\mathbf{K}^*_{\mathbf{SF}}$ both are constant filter matrix gates chosen for IF and for SF modes correspondingly. (1.11) assumes that $\mathbf{K}^*_{\mathbf{IF}}$ is used only during filter transitional process that can be terminated as soon as possible (for a short time) and then the steady filtering process can start and last as long as measurements vector $\mathbf{z}$ is available. This idea just reflects conventional engineering approach for linear control system design to extend system bandwidth at the transfer process period and to narrow it at the steady state work period. However, if it is clear that the steady state matrix-gate $\mathbf{K}^*_{SF}$ in (11) can be calculated using KBF formulas for $\mathbf{K}^*$ (12) (3-rd and 4-th). The question about how to determine the gate $\mathbf{K}^*_{IF}$ can be discussed additionally, but this gate should provide to the filter wider passband and bigger speed than $\mathbf{K}^*_{SF}$. Indeed, this matrix gate can be just designated using conventional engineering criteria such as stability margin, overshooting, and decaying time, considering filter characteristic polynomial

$$\Delta(s) = s\mathbf{I}\text{-}(\mathbf{F} + \mathbf{kH}) = s^n + a_1 s^{n\text{-}1} + \ldots\ldots + a_{n\text{-}1}s + a_0 \tag{14}$$

and arranging its coefficients a_i to provide to (14) desired roots. For example, standard coefficients for the multiple roots λ

$$\Delta(s) = (s + \lambda)^n \tag{15}$$

where λ determines system cut frequency (bandwidth) that is approximately inverse to its response time and is usually limited by the stability margin and static error under action of constant perturbations. Then matrix $\mathbf{K}^*_{SF}$ can be determined considering desired coefficients a_i (14) or the bandwidth λ and standard coefficients (15).

However, it is important to note that the reliable information about matrixes $\mathbf{Q}$ and $\mathbf{R}$ is usually not available almost at any stage of system development and operation. Especially, this is related to the matrix $\mathbf{Q}$. Hence, the question about the reliability of optimal filter gain $\mathbf{K}^*_{SF}$ also appears and can be discussed. In this case, when matrix $\mathbf{Q}$ is not available, it can be accepted as zero $\mathbf{Q} = \mathbf{0}$ and $\mathbf{P}^*$ is not optimal, but at some satisfactory level $\mathbf{P}^* = \mathbf{D}$. Then $\mathbf{K}^*_{SF}$ can be found from the equation

$$\mathbf{F}^*\mathbf{D} + \mathbf{P}^*\mathbf{D} + \mathbf{K}^*_{SF}\mathbf{H}\mathbf{R}^{-1}\mathbf{H}^T\mathbf{K}^{*T}_{SF} = \mathbf{0} \tag{16}$$

where $\mathbf{F}^* = \mathbf{F} - \mathbf{K}^*_{SF}\mathbf{H}$,

This equation is covariance equation for the filter estimation errors caused by the measuring noise $\mathbf{v}(t)$

$$\dot{\tilde{\mathbf{x}}} = \mathbf{F}^*\tilde{\mathbf{x}} - \mathbf{K}^*_{SF}\mathbf{H}\mathbf{v} \tag{17}$$

In fact, in practice the real physical processes $\mathbf{w}(t)$ and $\mathbf{v}(t)$ have more complex than Gaussian stationary white noise structure and hardly can be expressed by the matrixes $\mathbf{Q}$ and $\mathbf{R}$, assumed in KBF theory. However, this abstraction could be helpful for practical needs if some "appropriate" levels of $\mathbf{Q}$ and $\mathbf{R}$ that would be resulted in the solution, which is compatible with conventional engineering practice are taken to tune FBGM.

Practically, often there is no need to take care about the transfer mode of the KF and it can work permanently with the constant coefficients, determined by the formula below for the steady state [16]

$$\mathbf{K}^* = \mathbf{P}^*\mathbf{H}^T\mathbf{R}^{-1} \tag{18}$$

The steady state covariance matrix $\mathbf{P}^*$ as the solution of non-liner algebraic matrix Riccati equation below should be prematurely found

$$\mathbf{FP}^* + \mathbf{P}^*\mathbf{F}^T - \mathbf{P}^*\mathbf{H}^T\mathbf{R}^{-1}\mathbf{HP}^* + \mathbf{GQG}^T = \mathbf{0} \tag{19}$$

This filter can be used simultaneously as LTI negative feedback closed loop controller. Mathematically it can be derived by subtracting (7) from (3) and presented by the following equation

$$\dot{\tilde{\mathbf{x}}} = (\mathbf{F} - \mathbf{K}^*\mathbf{H})\tilde{\mathbf{x}} + \mathbf{KHV} + \mathbf{Gw} \tag{20}$$

These Eqs. (18)–(20) are used with example below to illustrate the KF analytical design approach with a simple LTI dynamical system of the second order.

3.1 Example

Let's consider a stable LTI, 2-nd order dynamic system

$$J\ddot{\alpha} + b\dot{\alpha} + c\alpha = M \tag{21}$$

where J is system moment of inertia, b is system damping coefficient, c is system rigidity coefficient, α is system angular position, $\dot{\alpha}$ is system angular velocity, M is external torque, applied to the system rotation part, having inertia J.

This simple dynamic Eq. (21) can approximately present many various real technical systems.

In particular, it can present the sensitive element of *the navigation accelerometer* (floating pendulum) with the electrical spring [16]. They were commonly used in electro-mechanic Inertial Navigation Systems (INS) in (70–80) – th. This device (system model (21)) will be considered further in this example.

The accelerometer measures external torque, applied to its pendulum by the inertia force $(F_i = -m_0 a)$ $M_i = -m_0 l a$, caused by its motion with acceleration a. In the steady state this torque is balanced by the accelerometer's spring $M_s = c\alpha_*$. Hence, pendulum deviation α_* allows determining the acceleration a

$$\alpha_* = -\frac{m_0 l}{c} a \tag{22}$$

To make the pendulum sensitive we need to eliminate the dry friction torque in its bearings M_{fr}. With this purpose the pendulum is usually put in the filled by the inert gas soldered chamber, floating in a viscous fluid in the accelerometer case. Then, the following inequality $M_{fr} \ll M_{i\,\min}$ takes place. The fluid also creates the viscous torque, damping the pendulum oscillations.

The following ratios describe the pendulum parameters: $J = m_0 l^2$ (m_0 is pendulum mass, l is its length), b is pendulum floating chamber viscous friction coefficient, $c = m_0 g l$ is pendulum rigidity coefficient, creating by the pendulum weight. Substituting these parameters in (21) and dividing it by $J = m_0 l^2$, we can represent it as follows

$$\ddot{\alpha} + 2d_0\omega_0\dot{\alpha} + \omega_0^2\alpha = -m \tag{23}$$

where $2d\omega_0 = \frac{b}{J} = \frac{b}{m_0 l^2}$, $\omega_0^2 = \frac{c}{J} = \frac{g}{l}$, $m = \frac{M}{J} = \frac{m_0 l}{m_0 l^2} a = \frac{1}{l} a$. Hence,

$\omega_0 = \sqrt{\frac{g}{l}}$ is system natural frequency, $d_0 = \frac{b}{2m_0 l^2 \sqrt{\frac{g}{l}}} = \frac{b}{2m_0 \sqrt{gl^3}}$ is system specific damping coefficient.

If measured acceleration a is constant ($a = const$), then in the steady state ($t \to \infty$) from (23) follows that

$$\alpha(\infty) = \alpha_* = -\frac{1}{g}a \tag{24}$$

In further consideration we will assume that the devise angle α is measured by a kind of the electric sensor, providing the electric output constant voltage as follows

$$U_a = -k_a \alpha = a \tag{25}$$

where $k_a = -g$ is the pendulum scale coefficient.

In our case (23) in the canonical form of the 2-nd order differential equation dynamic unit presents uncontrolled system (a plant).

The purpose of further design is to find required for the pendulum (23) control that would provide to it (accelerometer) better dynamics. Usually the viscous fluid provides for the floating pendulum a big damping coefficient d_0, however the "gravity spring" has not enough rigidity to have for the device required natural frequency ($\omega_0 \leq \omega$). Therefore, we will develop for the pendulum additional (to the gravity torque) negative feedback electric positional torque ("electric spring") that will increase original pendulum natural frequency ω_0 to the desired value of ω. Schematically such an accelerometer is presented in **Figure 2**.

Two approaches are considered below: A-Empirical design (based on engineer experience and the continuity with the conventional design) and B-Analytical design (based on the suboptimal KF (FBGM) used as a linear controller).

3.2 Empiric approach

Eq. (22) can be presented in Laplas transformation form as follows

$$\left(s^2 + 2d_0\omega_0 s + \omega_0^2\right)\alpha(s) = -\frac{1}{l}a(s) \tag{26}$$

where s is Laplas operator, $\alpha(s)$ and $a(s)$ are the Laplas transformations of the output pendulum angle and its input acceleration correspondingly.

Let us apply in inverse to this pendulum deviation direction the control torque M_c that physically would provided by a special electric motor (in the" braking mode"), installed on the pendulum rotation shaft and controlled by the voltage from the sensor of pendulum deviation angle α through the electronic amplifier. This negative feedback is the accelerometer "electric spring" (see **Figure 2**)

Mathematically, this specific control torque in (22) can be presented as follows

$$m_c = \frac{M_c}{J} = -k_s(\alpha + \Delta\alpha) - \varepsilon(\dot{\alpha} + \Delta\dot{\alpha}) \tag{27}$$

where m_c is specific control torque applied by the "electric spring" (27), $k_s = \frac{k_a k_{am} k_{M_c}}{J}$ is the electric spring rigidity coefficient, k_a is angle α scale coefficient, k_{am} is

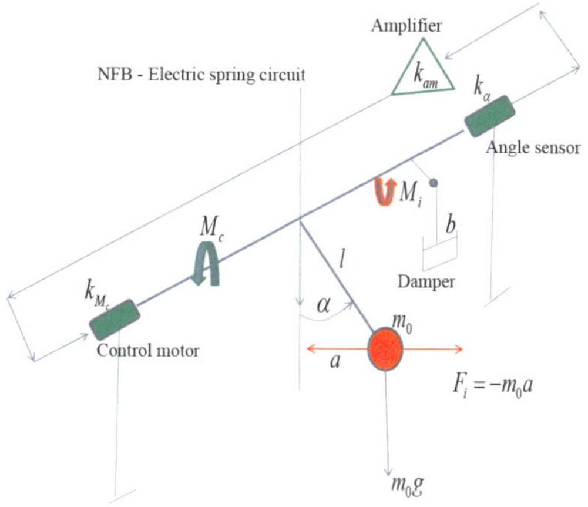

Figure 2.
Scheme of the accelerometer with the electric spring.

amplifier scale coefficient, k_{M_c} is control motor scale coefficient, $\Delta\alpha$ is angular sensor α measured error.

Substituting (27), in (26), we can rewrite it as follows

$$\left[s^2 + (2d\omega + \delta)s + \omega^2\right]\alpha(s) = \omega_0^2 n(s) + (\delta s + k_s)\Delta\alpha(s) \tag{28}$$

where $\omega^2 = \omega_0^2 + k_s = k_s\left(1 + \frac{\omega_0^2}{k_s}\right)$, $d = \frac{b}{2J\omega} = d_0\frac{\omega_0}{\omega}$, $n = \frac{a}{g}$ is measured input overload, δ is a small additional damping coefficient, created by the counter electromotive force in the winding of the motor-torque in the electric spring. It is neglected in further consideration $\delta \approx 0$

The output voltage can be determined as in (25), measuring the angle α as $U = -k_a\alpha$. The scale coefficient of the accelerometer is determining by the following formula

$$k_a = -l\omega^2 = -l\left(\omega_0^2 + k_s\right) \tag{29}$$

Usually, the ratio $\varepsilon = \frac{\sqrt{k_s}}{\omega_0}$, is much bigger than one $\varepsilon \gg 1$ or $\frac{1}{\varepsilon^2} \ll 1$, then for the accelerometer dynamic parameters takes place following approximate formulas

$$\omega \approx \sqrt{k_s} = \varepsilon\omega_0 \tag{30}$$

$$d \approx \frac{d_0}{\varepsilon} \tag{31}$$

The coefficient $\varepsilon = \frac{\omega}{\omega_0}$ is accelerometer bandwidth increasing coefficient. We can see from (30) to (31) that the electric spring will increase the accelerometer natural frequency and in the same ratio decrease its damping coefficient. That is why original damping coefficient for the devise has to be chosen much bigger then is usually required for the 2-nd order dynamic unit $d \gg 0.707$.

Generalizing presented above results of the empirical design approach to the electric spring accelerometer, we can conclude as a rule of thumb that using negative positional feedback increases dynamic system bandwidth, however decreases its damping coefficient.

Numerical data and simulation results for the approach A are presented below

3.3 Free pendulum

Pendulum parameters

$$m_0 = 5 \cdot 10^{-3} Kg,\ l = 0.02m,\ J = m_0 l^2 = 2 \cdot 10^{-6} kgm^2,\ b = \ 6.295 \cdot 10^{-4} \mathrm{Nm/rad/s},$$
$$c = m_0 gl = 9.8 \cdot 10^{-4} Nm/rad,\ k_{a0} = -9.8 Vs^2/m$$

Pendulum standard dynamics coefficients can be calculated as following

$$\omega_0 = 22.136\ \mathrm{s}^{-1}\ (f_0 = 3.523\ \mathrm{Hz},\ T_0 = 0.0452s),\ d = 7.1$$

Simulink simulation scheme is presented in **Figure 3**
Step-response of this pendulum is presented in **Figure 4**.
This pendulum with electric spring was simulated with similar Simulink scheme and methodology. Simulation results are presented below.

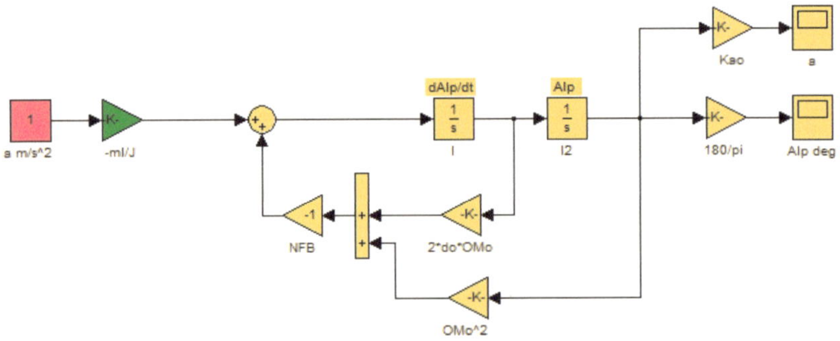

Figure 3.
Simulink block-diagram of the floating pendulum-system model.

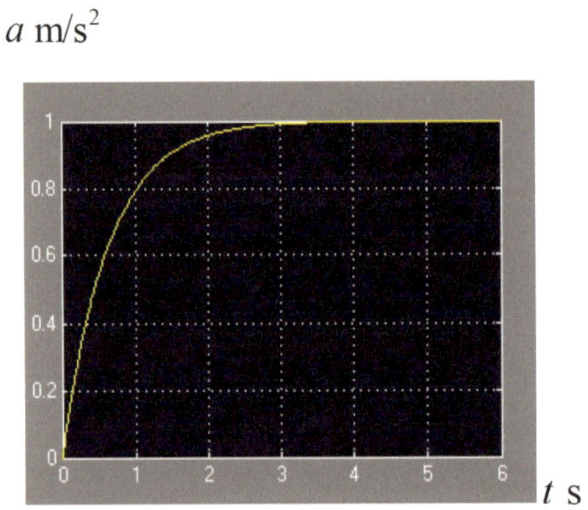

Figure 4.
Step-response of the pendulum to its input acceleration $a = 1$ m/s.

3.4 Pendulum with the electric spring-accelerometer

3.4.1 Desired coefficients

Let us assume that we want to increase the device natural frequency that characterizes its bandwidth in ten times to have the ratio $\lambda = \frac{f}{f_0} = 10$. Then the accelerometer will have the following dynamics parameters: $\omega = 221.36$ rad/s, $(f = 35.24Hz)$ $d = 0.71$. The electric spring specific rigidity coefficient is $k_s = \omega^2 - \omega_0^2 = 4.851 \cdot 10^4 \text{rad/s}^2$ $(k_s = \frac{K_s}{J})$ and the scale coefficient is $k_a = -980\ Vs^2/m$.

The accelerometer Simulink simulation scheme is presented in **Figure 5**.

There were introduced measurement random errors $\Delta\alpha$ (electronic noise) as a Band-limited Gaussian White Noise with $\sigma_{\Delta\alpha} = 0.01^0$ and $\Delta t = 0.001s$; $f_m' = \frac{1}{\Delta t} = 1kHz$ (see White Noise generator block in pink in **Figure 5**). This measured noise is presented in **Figure 6**.

Step-response of the accelerometer deviation angle α is presented in **Figure 7**.

The filtering capability of this device we can evaluate by finding the STD of its output angle σ_α, caused by the electronic noise $N = \sigma_{\Delta\alpha}^2 \Delta t$ of the electric spring. This STD can be calculated with formula, following from [24]. Using this equation for the steady state, we can found that

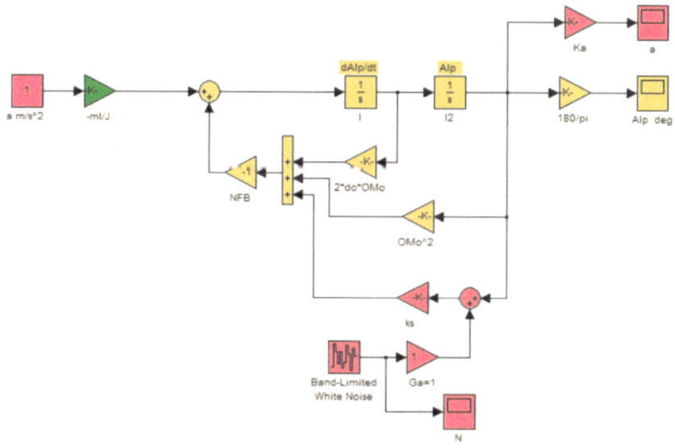

Figure 5.
Simulink block-diagram of the Electric Spring Accelerometer.

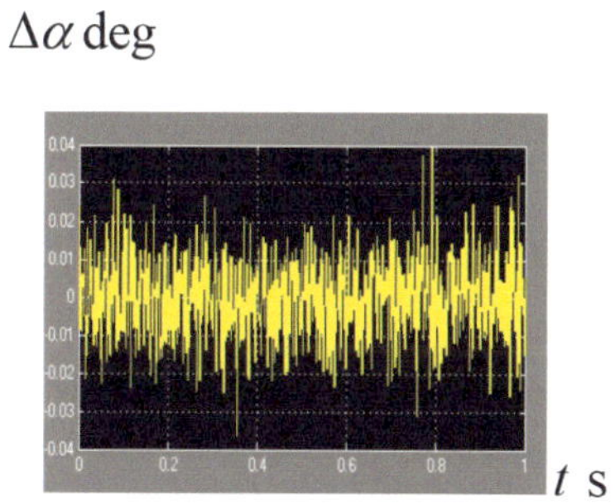

Figure 6.
Measurement random electronic Noise: $\sigma_{\Delta\alpha} = 0.01^\circ$, $\Delta t = 0.001$ s.

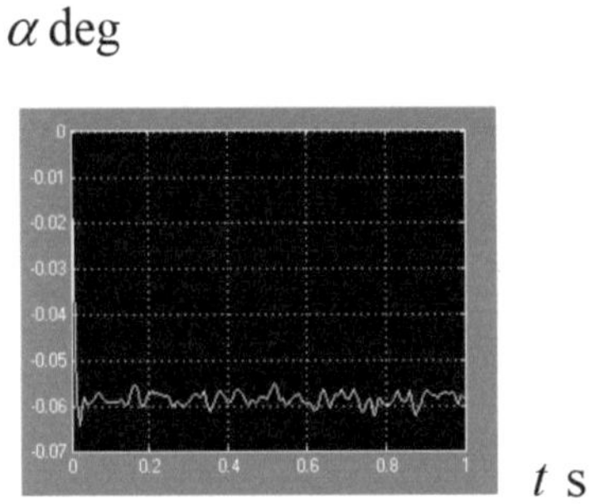

Figure 7.
Step-response $\alpha(t)$ of the accelerometer to its input acceleration $a = 1$ m/s, $\alpha_*(t \to \infty) = -0.058^\circ$, $\sigma_\alpha \approx 0.005^\circ$.

$$\sigma_\alpha = \sqrt{\frac{\pi f}{d f_N}} \sigma_{\Delta\alpha} \tag{32}$$

where f is accelerometer natural frequency in *Hz*, d is accelerometer specific damping coefficient, $f_N = \frac{1}{\Delta t}$ is the range of the frequency in *Hz*, where the noise can be practically considered as a "white", Δt is the noise sampling time. Using (32) we can calculate that in our example the ratio $r = \frac{\sigma_\alpha}{\sigma_{\Delta\alpha}} = 0.395$. That is, the filtering capability of the accelerometer cuts its random electronic noise approximately in 2.5 times.

As we can see from the above results, the empiric approach A allowed us for getting a quite satisfactory accelerometer dynamics.

3.4.2 Analytical design approach

Let us introduce a simplest 2-nd order unit, to use it as the system (future accelerometer sensor) model

$$\begin{cases} \dot{x}_1 = w \\ \dot{x}_2 = x_1 \end{cases} \tag{33}$$

where w is system disturbance, x_1 and x_2 are system state vector variables.

Let us assume that the state vector variable x_2 is measured

$$z = x_2 + v \tag{34}$$

where v is an additive measured error.

Let us also assume that w and v are Gauss white noises, having spectral densities q and r, respectively and find the optimal KF, minimizing the STD of estimated system state vector random errors [14].

Then KF for (33) and (34) can be written as follows

$$\begin{cases} \dot{\hat{x}}_1 = k_{12}(z_2 - \hat{x}_2) \\ \dot{\hat{x}}_2 = \hat{x}_1 + k_{22}(z_2 - \hat{x}_2) \end{cases} \tag{35}$$

The control gaits in (35) can be calculated with formulas

$$\begin{cases} k_{12} = \dfrac{P_{12}^*}{r} \\ k_{22} = \dfrac{P_{22}^*}{r} \end{cases} \tag{36}$$

where P_{ij}^* are elements of the steady state covariance matrix P^* that can be found solving the algebraic equation Riccati [18] for (33)–(35).

Considered above case allows for the analytical solution of [18] that is as follows [9]

$$\begin{cases} P_{12}^* = \sqrt{q_1 r} = r\sqrt{\xi}, \\ P_{22}^* = \sqrt{2rP_{12}^*} = r\sqrt{2\sqrt{\xi}}, \\ P_{11}^* - \dfrac{1}{r}P_{12}^* P_{22}^* = r\sqrt{2\xi\sqrt{\xi}}, \end{cases} \tag{37}$$

where $\xi = \frac{q}{r}$ is ratio of system exiting noise to measured error noise spectral densities and has meaning of "*filterability*". Substituting P_{12}^* and P_{22}^* from (37) in (36) we can determine the filter coefficients with formulas

$$\begin{cases} k_{12} = \sqrt{\xi}, \\ k_{22} = \sqrt{2\xi\sqrt{\xi}}, \end{cases} \tag{38}$$

The KF (35) can be used as the closed loop negative feedback controller [9]. This closed loop system can be presented if we subtract (35) from (33). It will be presented by the following equations

$$\begin{cases} \dot{\tilde{x}}_1 = -k_{12}\tilde{x}_2 + k_{12}v + w \\ \dot{\tilde{x}}_2 = \tilde{x}_1 - k_{22}\tilde{x}_2 + k_{22}v \end{cases} \tag{39}$$

where $\tilde{x}_1 = \hat{x}_1 - x_1,\ \tilde{x}_2 = \hat{x}_2 - x_2$ are errors of control of the closed loop system (39). This system can be converted to the following form

$$\ddot{\alpha} + k_{22}\dot{\alpha} + k_{12}\alpha = w + k_{22}\dot{v} + k_{12}v \tag{40}$$

Let us introduce the following designations $k_{22} = 2d\omega,\ k_{12} = \omega^2$, then (40) can be rewritten in the following Laplas transformation form

$$\left(s^2 + 2d\omega s + \omega^2\right)\alpha = w(s) + \left(2d\omega s + \omega^2\right)v(s) \tag{41}$$

Where $\omega = \sqrt{k_{12}} = \sqrt[4]{\xi}$ is filter natural frequency, $d = \frac{k_{22}}{2\sqrt{k_{12}}} = \frac{\sqrt{2}}{2} = 0.707$ is filter specific damping coefficient.

Let us put in (40) that $w = -\frac{1}{l}a = -\omega_0^2 n$, where $n = \frac{a}{g}$ is overload, $\omega_0^2 = \frac{g}{l}$, than (41) can be rewritten as following

$$\left(s^2 + 2d\omega s + \omega^2\right)\alpha = -\omega_0^2 n(s) + \left(2d\omega s + \omega^2\right)v(s) \tag{42}$$

Now this equation (42) may represent the accelerometer with the electric spring, like (28). But, unlikely (28) and (41) presumes that all system damping is created electrically (by the negative feedback compensation torque-spring). However, the similarity of the transfer functions (output/input) of (28) and (42) allows us to note some generic futures of the optimal 2-nd order LTI unit. It was already found that optimal specific damping coefficient must be 0.707 ($d = 0.707$) and this fact doesn't depend on w and v statistic characteristics. Let us determine the ratio ξ for (41)

$$\xi = \frac{q}{r} = \frac{\omega_0^4 \sigma_n^2 \Delta t_a}{\sigma_{\Delta\alpha}^2 \Delta t_{\Delta\alpha}} \tag{43}$$

Then the natural frequency for (42) will be determined by the formula

$$\omega = \sqrt[4]{\xi} = \omega_0 \sqrt{\frac{\sigma_n}{\sigma_{\Delta\alpha}} \sqrt{\frac{\Delta t_a}{\Delta t_{\Delta\alpha}}}} \tag{44}$$

Using analytical design approach B, we hardly can count on the availability of reliable statistic characteristics of w and v. Rather, we can set them empirically from the standpoint of common sense. For example, let's we want to get the same system as using approach A. In another words we want to have the same ratio $\lambda = \frac{\omega}{\omega_0} = 10$, as in the example A, then we must assume that takes place the following ratio $\sqrt{\frac{\sigma_n}{\sigma_{\Delta\alpha}} \sqrt{\frac{\Delta t_a}{\Delta t_{\Delta\alpha}}}} = 10$ or $\frac{\sigma_n \sqrt{\Delta t_a}}{\sigma_{\Delta\alpha} \sqrt{\Delta t_{\Delta\alpha}}} = 10^2$. Then this ratio can be rewritten as follows

$$\frac{\sigma_n}{\sigma_{\Delta\alpha}} = 10^2 \sqrt{\frac{f_{\Delta\alpha}}{f_a}} \tag{45}$$

Let's assume, as using approach A, that $f_{\Delta\alpha} = 10^3 Hz$ and accept that $f_a \geq f$ ($f = 35.24 Hz$), $f_a = 100 Hz$. Then, as follows from (47), for the equivalency of both approaches, namely A and B, should take place the following equality $\sigma_n = 316 \sigma_{\Delta\alpha}$ (where $\sigma_{\Delta\alpha} = 0.01° = 0.017 rad$). Hence, $\sigma_n = 5.5$ ($\sigma_a = 5.5g = 54 m/s^2$).

Simulation result of extracting by the KF filter (39) (accelerometer) a noise-like signal from the white noise is presented in **Figure 8**.

We can see that synthesized by using KF accelerometer is quite efficient to measure not only step-like input accelerations (as it was showed by the approach A), but the white noise-like accelerations as well. If considered above numerical data are accepted, then both approaches led to the same system (accelerometer) dynamics.

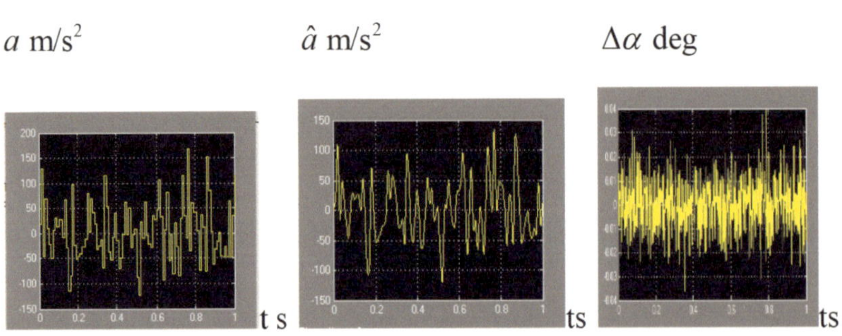

Figure 8.
Extracting by the synthesized KF accelerometer the noise-like signal from the measured white noise. a) Input noise: $\sigma_a = 54\ m/s^2$, $\Delta t_a = 0.01s$ *b) Estimates* $\hat{a}$ *c) Measured noise:* $\sigma_{\Delta\alpha} = 0.01°$, $\Delta t_{\Delta\alpha} = 0.001s$.

Presented above KF results are rather qualitative, assuming a dry accelerometer (without damping by the viscous fill fluid), where the damping was created electrically (in the negative positional feedback). Basically, they can be appropriately converted and applied to the practical design. However, more accurate KF synthesis for the accelerometer (23) can be performed, considering the following model

$$\begin{aligned} \dot{x}_1 &= -2d_0\omega_0 x_1 - \omega_0^2 x_2 - \omega_0^2 w \\ \dot{x}_2 &= x_1 \\ z &= x_2 + v \end{aligned} \tag{46}$$

where $x_1 = \dot{\alpha}, x_2 = \alpha, v = \Delta\alpha, w = \frac{a}{g}$.

$$\begin{aligned} \dot{\hat{x}}_1 &= -2d_0\omega_0\hat{x}_1 - \omega_0^2\hat{x}_2 + k_{12}(z - \hat{x}_2) \\ \dot{\hat{x}}_2 &= \hat{x}_1 + k_{22}(z - \hat{x}_2) \end{aligned} \tag{47}$$

where $k_{12} = \frac{P_{12}^*}{r}$, $k_{22} = \frac{P_{22}^*}{r}$

Using this model, the readers can analyze presented example more precisely. The author leaves this option for readers' independent exercise

4. Conclusion

In the example above the KF was used for the analytical design to find optimal dynamic characteristics for the navigation accelerometer. Optimal tuning for the devise coefficients were found, which can be easily implemented in the accelerometer hardware, using analogue elements (floating pendulum, precision bearings, electric angle sensor, electronic amplifier and torque motor). Comparing the results with the conventional design we can conclude that the design is a stable and robust, as well as for the conventionally designed accelerometer. Generalizing, we can conclude that using the KF for the analytical design in engineer applications leads to quite realistic results that can be verified with conventional solutions. It can also be noted, that even using the analytical design, the choice of the appropriate values for the KF matrixes **Q** and **R** is rather based on the developer experience and intuition than on real statistic characteristics of the processes $\mathbf{W}(\mathbf{t})$ and $\mathbf{V}(\mathbf{t})$, which are unlikely be available to use them.

Acknowledgements

The author dedicates this Chapter to his mentor colonel, professor of Moscow Aviation Institute and Zhukovsky Air Force Engineering Academy V. P Selesnev, who introduced him in Kalman Filtering applications.

Author details

Yuri V. Kim
David Florida Laboratory (DFL), Ottawa; Canadian Space Agency (CSA), St. Hubert, Gatineau, Quebec, Canada

*Address all correspondence to: yurikim@hotmail.ca

References

[1] Kalman RE. A new approach to linear filtering and prediction problems. ASME Journal of Basic Engineering. 1960;**82**(1): 34-45

[2] Kalman RE, Bucy RS. New results in linear filtering and prediction theory. ASME Journal of Basic Engineering. 1961;**80**(1):193-196

[3] Kim YV. An approach to suboptimal filtering in applied problems of information processing. USSR Journal of Science Academy. 1990;**1**(1):109-123

[4] Kim YV, Kobzov PP. Optimal filtering of a polynomial signal. USSR Journal of Science Academy. 1991;**2**(1): 120-133

[5] Kim YV, Nazarov AB. Synthesis of optimal dynamic characteristics for a single axis Inertial Navigation System in a steady state. LITMO. 1975; **18**:80-85

[6] Kim YV, Goncharenko G. On an approach to observability analysis in INS correction problems. Orientation and Navigation. 1981;**11**:25-28

[7] Kim YV, Ovseevich AI, Reshetnyak YN. Comparison of stochastic and guaranteed approaches to the estimation of the state of dynamic system. USSR Journal of Science Academy. 1992;**2**(1):87-94

[8] Kim YV. Kalman filter decomposition in the time domain using observabiliy index. In: International Conference Proceedings. New York: Curran Associates Inc; 2008

[9] Kim UV. Kalman filter and satellite attitude control system analytical design. International Journal of Space Science and Engineering. 2020;**6**(1):94-95

[10] Zarach P, Missof H. Fundamentals of Kalman Filtering: Practical Approach. Reston, VA: American Institute of Aeronautics and Astronautics, Inc.; 2000

[11] Vukovich G, Kim Y. The Kalman Filter as controller: Application to satellite formation flying problem. International Journal of Space Science and Engineering. **3**(2):148-170

[12] Bryson AE, Yu-Chi HJ. Applied Optimal Control. Levittown, PA: Taylor & Francis; 1975

[13] Chernousko FL, Kolmanovskiy VB. Optimal Control under Random Disturbances. Russia: Nauka; 1978

[14] Gelb A. Applied Optimal Estimation. Cambridge, MA: The M.I.T. Press; 1974

[15] Krasovsky A. Automatic Flight Control systems and Analytical Design. Russia; 1973

[16] Seleznev VP. Navigation Devices. Russia; M. Mashinostroenie; 1974

[17] Bellman R. Dynamic Programming. Princeton NJ: Princeton University Press; 1957

[18] Kwakernaak M, Sivan H. Linear Optimal Control Systems. Russia: Mir; 1972. pp. 83-88

[19] MathWorks, Inc. Kalman Filtering-Matlab and Simulink, Documentation. The MathWorks. Inc; https://www.mathworks.com › control, visited Aug. 2022

[20] Pontriagin L, Boltiasky V, Gamkrelidze R, Mischenko E. Mathematical Theory of Optimal Processes. Nauka; 1976

[21] Aarenstrap R. Managing Model-Based Design. Natick, MA: Math Works, Inc; 2015

[22] Letov A. Flight Dynamics and Control. Russia: Nauka; 1979

[23] MathWorks I. Training Course, Adopting Model based Design. Natick, MA: The MathWorks. Inc; 2005

[24] Kim YV, Ovseevich AI, Reshetnyak YN. Comparison of stochastic and guaranteed approaches to the estimation of the state of dynamic system. USSR Journal of Science Academy. New York: Scripta Technica Inc. (Rus., Engl. Transl). 1992;2:87–94

Chapter 2

Extended Kalman Filter for a Monitoring System of the Guyed Towers

Alexandre Schalch Mendes, Pablo Siqueira Meirelles, Janito Vaqueiro Ferreira and Eduardo Rodrigues de Lima

Abstract

The Kalman filter is used in a wide range of systems. Due to its efficiency to estimate the state variables in structures and mechanisms, its usage is already well known in Control Systems and in addition to smooth measured and unmeasured signals, it allows the sensor fusion technique to consider the best characteristics of each type of sensor. In this Chapter, we will present the application of the Extended Kalman Filter (EKF) to predict the movement of the upper top center of a guyed tower of a power line, based on its dynamic model and considering the indirect measurement of the force in the stay cables through accelerometers. In a mockup tower, built in Dynamic Testing Laboratory (LabEDin) at University of Campinas (UNICAMP), it is possible to apply external loads and simulate failures, such as degradation of the stay cable foundation. Variations in the cable forces are used as inputs in the EKF algorithm and as estimate, we obtain the amplitudes and directions of the tower's top movements, which will be considered to predict the health of its structure, indicating the need for maintenance intervention. All the theories considered in this proposed methodology are validated with the experiments carried out on the mockup tower in the laboratory.

Keywords: extended Kalman filter, sensor fusion, IoT, guyed towers, transmission power lines

1. Introduction

The objective of this Chapter is to present a practical application of the Kalman filter considering a dynamic model of a steel tower structure, which is not usual for a Bayesian filtering utilization. Usually this filtering technique is considered to estimate time-dependent physical quantities, such as in Global Positioning Systems (GPS), target tracking, satellite control, etc.

The origin of Bayesian analysis belongs to the field of optimal linear filtering and the concept of building mathematically optimal recursive estimators, was first presented for linear systems due to their mathematical simplicity [1]. The success of the Bayesian filtering applied to engineering problems, started with an article presented in 1960 by the Hungarian-American mathematician and engineer, *Rudolf Emil Kálmán*. He developed a mathematical method for linear filtering which does not requires powerful computing machinery and several applications of his method led to further application of the named Kalman filter in Apollo program, NASA Space Shuttle, Navy submarines, in Unmanned Aerospace Vehicles (UAV) and also in weapons, such as cruise missiles.

In our studied case, the guyed towers of a transmission power line need to be inspected along the year, to check the inclination and the forces at the stay cables to assure its structural integrity. In order to reduce the preventive maintenance costs, there is a proposal for a remote monitoring system to estimate the position of the upper part of the tower based on the indirect measurements of stay cable forces. The natural frequencies of the cables can be determined from time-domain accelerations, obtained from accelerometers connected to the stay cables. Based on an existing relation between the cable force and its natural frequency, it is possible to calculate the instantaneous force at every acceleration measurement. As redundancy, the signal from an inclinometer will be fused into the algorithm to provide the direction of the tower movement. Another important characteristic of the Kalman filter to be highlighted, is that in addition to fuse different types of sensors, it is possible to estimate variables which are not being measured directly.

The proposed monitoring system will transfer (via RF) the accelerations measured in time-domain from the leaf nodes positioned in each stay cables, to a central router in the tower. These signals will be transferred to a border router responsible to organize the acquisitions of several towers to a main server, where the post processing will be carried out. The EKF algorithm and all calculations are performed in a MATLAB® (Mathworks, EUA) code developed at the University of Campinas (UNICAMP).

2. Kalman filter concepts

When any signal is collected through a sensor, generally it presents some level of noise which can directly affect the evaluation of the measured quantity. The usage of the Kalman filter, in addition to smooth the measured signal, allows to estimate quantities which are not being measured directly. And as also mentioned before, Kalman filtering presents the possibility to fuse signals from different sensor types to achieve the desired objectives [2]. Let us consider an instant $\boldsymbol{k}$ of the measured signal. At that time, the prediction of the estimated signal $\hat{\boldsymbol{x}}_{\bar{k}}$ can be calculated considering the Eq. (1), as follow:

$$\hat{x}_{\bar{k}} = A\hat{x}_{k-1} \tag{1}$$

And the error covariance for the predicted instant $\bar{\boldsymbol{k}}$ will be calculated as shown in Eq. (2).

$$P_{\bar{k}} = AP_{k-1}A^T + Q \tag{2}$$

Where $\boldsymbol{A}$ is the state transition matrix that contains the characteristics of the physical model with dimensions (n x n) and $\boldsymbol{Q}$ is the noise covariance matrix of the state transition (n x n). The next step within the algorithm will be the determination of Kalman gain, which will be recursively calculated within the code for the entire signal length. The Kalman gain can be obtained via Eq. (3).

$$K_k = P_{\bar{k}} H^T \left(H P_{\bar{k}} H^T + R\right)^{-1} \tag{3}$$

Where $\boldsymbol{R}$ is the covariance matrix of the measurement noise (m x m) and $\boldsymbol{H}$ is the state matrix for the measurement (m x n). With these data, the estimated values and the covariance of the signal error will be finally calculated. The estimate signal $\hat{\boldsymbol{x}}$ (n x 1) can be determined by Eq. (4) from the measured signal $\boldsymbol{z}$ (m x 1).

$$\hat{x}_k = \hat{x}_{\bar{k}} + K_k \left(z_k - H \hat{x}_{\bar{k}}\right) \tag{4}$$

And the error covariance, through Eq. (5):

$$P_k = P_{\bar{k}} - K_k H P_{\bar{k}} \tag{5}$$

Behind this formulation, there are sophisticated statistical concepts that allow the Kalman filter not only to smooth the signals, but also to estimate the values of different quantities in regions of the structure where we do not have access for instrumentation. It is important to highlight that the matrix $\boldsymbol{H}$ is the responsible to correlate measurements and estimations of different quantities within the algorithm.

So far, the application of the Kalman filter in linear systems has been presented. This methodology cannot be applied when there are time-dependent variations in the state transition matrix $\boldsymbol{A}$ and/or in the state matrix for the measurement $\boldsymbol{H}$, i.e. nonlinear systems. In these cases, the Extended Kalman filter is used to estimate the system outputs. In a simple manner, the next equations present the difference between the conventional (or linear) Kalman filter and the Extended Kalman filter that shall be considered for nonlinear systems.

Thus, the algorithm for the application of the Extended Kalman filter, shall be able to introduce the non-linearities presented by Eqs. (6) and (7). **Figure 1** illustrates the diagram used to develop the computational code considering the Extended Kalman filter for the tower model. It is possible to consider a linearization technique through the Jacobian matrix, as follow:

$$Ax_k \Rightarrow f(x_k); A \equiv \frac{\partial f}{\partial x} |\hat{x}_k \tag{6}$$

$$Hx_k \Rightarrow h(x_k); H \equiv \frac{\partial h}{\partial x} |\hat{x}_{\bar{k}} \tag{7}$$

The objective of this Chapter is to give a brief explanation about Kalman filter theory and focus on the application of the algorithm in an engineering problem. There are several good references for a deeper explanation about linear and nonlinear Kalman filtering methodology, such as presented in [3, 4].

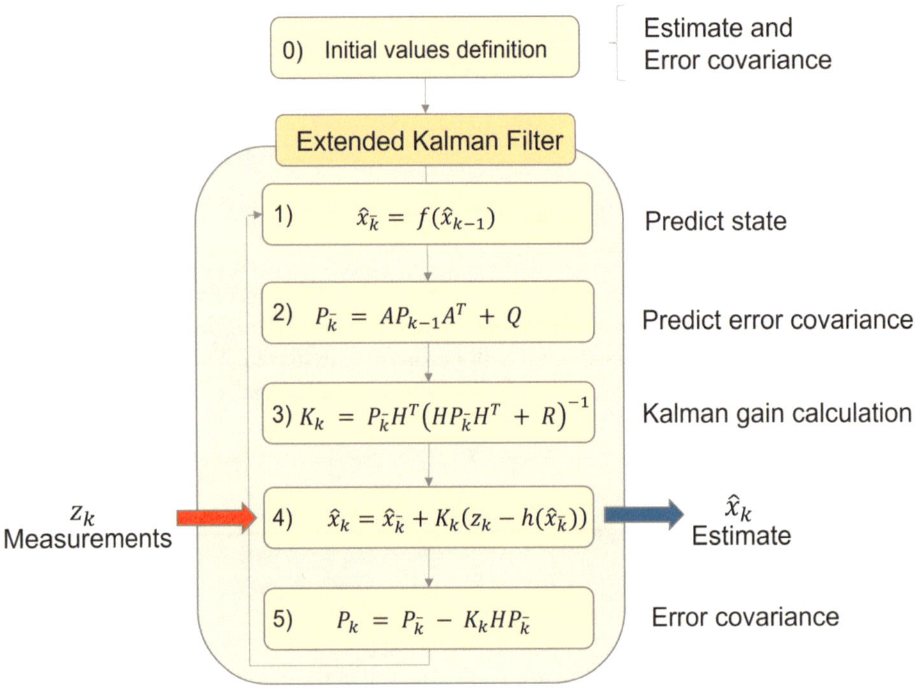

Figure 1.
Representation of the extended Kalman filter algorithm.

3. Determination of the natural frequencies of the stay cables

Based on the proposed methodology to estimate the displacement of the tower using EKF, it is first necessary to define a relation between the actual stay cable force and the natural frequencies of this component. This mathematical relation is based on the equation of the uni-dimensional wave propagation in a cable fixed in both extremities. In [5], Steidel proposed an equation for cables with different lengths and cross sections. The steel cables used in the mockup tower have a core fiber, which presents a nonlinear behavior and different dynamic responses when compared to the all-metal cable of the theory presented in [5]. Thus, a verification of these cables was proposed in the Dynamic Testing Laboratory (LabEDin) at UNICAMP, to determine the actual relationship between cable force and natural frequencies.

Figure 2 shows the devices used in LabEDin to apply the forces and determine the natural frequencies of the cable with the same equivalent length of the stay cables of the mockup tower. The load was applied in the cable through a hydraulic cylinder and the acceleration signals were measured from both sensors at the positions of $L_e/2$ and $L_e/4$ of the cable equivalent length $L_e = 4.3m$. It was observed that there are no significant differences to determine the natural frequencies from the PSDs, with or without impacts on the cable.

The experiment consists of applying constant loads from 1000 to 2600 N with an increment of 100 N and determine the first three natural frequencies of the cable from the measured signals. The variation of the cable length was also taken into account, as presented in [6], to define the equation that relates the natural frequency as a function of the cable force. The Eq. (8) shows this relationship for the 1/4″ diameter steel wire ropes used to clamp the tower.

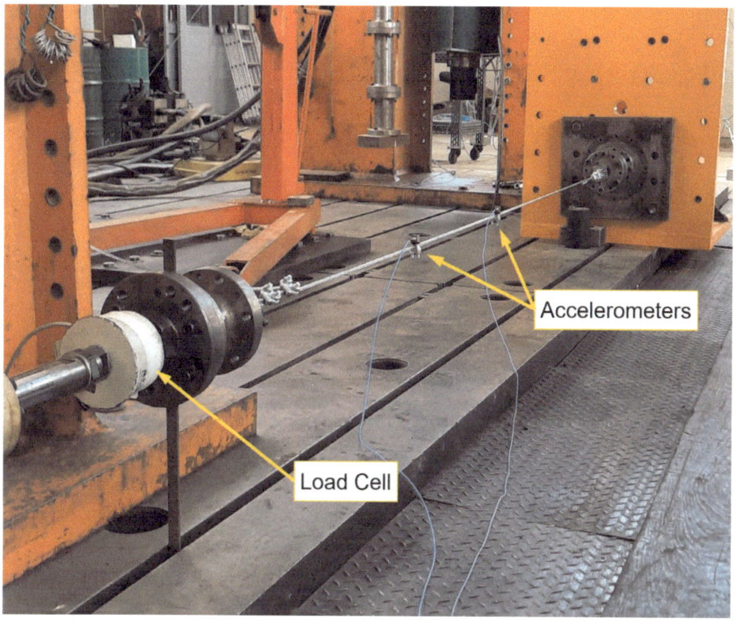

Figure 2.
Device for testing the stay cable of the mockup tower.

$$\omega_n = \frac{0.3027n\sqrt{\frac{F}{\mu}}}{{L_e}^{0.718}} \tag{8}$$

whcrc:

ω_n	Natural frequency of the cable [Hz]
n	Mode of vibration [−]
F	Tensile force on the cable [N]
μ	Linear mass density [kg/m]
L_e	Equivalent cable length [m]

Figure 3 presents the measured natural frequencies of the first three vibration modes of the cable, compared with the values calculated using Eq. (8) as a function of the applied force. It is possible to notice that an accurate approximation of the calculated values was achieved using the equation, which allows considering this procedure to determine the cable forces via accelerometers as input to the EKF algorithm.

4. Dynamic model of the guyed tower

A successful application of the Extended Kalman filter, depends on a well defined system model and the reliability of the outputs is directly related to the ability of the model to reproduce the physics of the actual mechanical structure. The dynamic

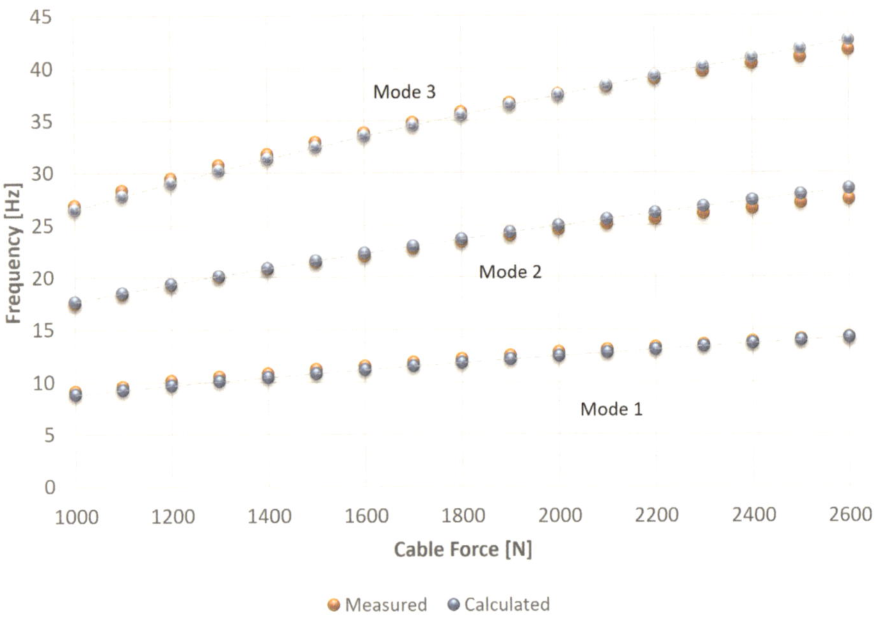

Figure 3.
Comparison between the measured and calculated natural frequencies of the first three modes, as a function of the cable load.

model of the guyed tower was developed to provide fast responses as function of the force variation on the stay cables.

Figure 4 shows the picture of the tower assembled in LabEDin, indicating the position of the inclinometer, accelerometers and load cells. There is one accelerometer (triaxial) and one load cell (S-type) per stay cable. The load cells are used just to validate the calculation of the cable force using Eq. (8). The structure of the mockup tower was designed based on a 1:5 scale of a power line tower with 22.5 meters height.

The dynamic equations of the tower model were defined considering the Lagrange function [7], as presented in Eq. (9). The angles $\boldsymbol{\theta_x}, \boldsymbol{\theta_y}, \boldsymbol{\theta_z}$ at the bottom point of the tower are considered as generalized coordinates, with $\boldsymbol{i} = \mathbf{3}$, defining the number of degrees of freedom (DOF) of the model:

$$\frac{\mathrm{d}}{\mathrm{d}t}\left(\frac{\partial \mathcal{L}}{\partial \dot{q}_i}\right) - \frac{\partial \mathcal{L}}{\partial q_i} = Q_i \tag{9}$$

with:

$$\mathcal{L} = T - U = T_v + T_\omega - \left(U_g + U_K + U_{K_t}\right) \tag{10}$$

Figure 5 shows the proposed model of the guyed tower. In this figure are presented the main dimensions of the mockup tower and the global system of coordinates. The structure can rotates freely in the three direction at the pivot point represented by the green circle.

The kinetic energy shown in Eq. (10) is determined as follows:

$$T_v + T_\omega = \frac{1}{2} m \; v_{cg}^2 + \frac{1}{2}\left(I_{xx} \; \dot{\theta}_x^2 + I_{yy} \; \dot{\theta}_y^2 + I_{zz} \; \dot{\theta}_z^2\right) \tag{11}$$

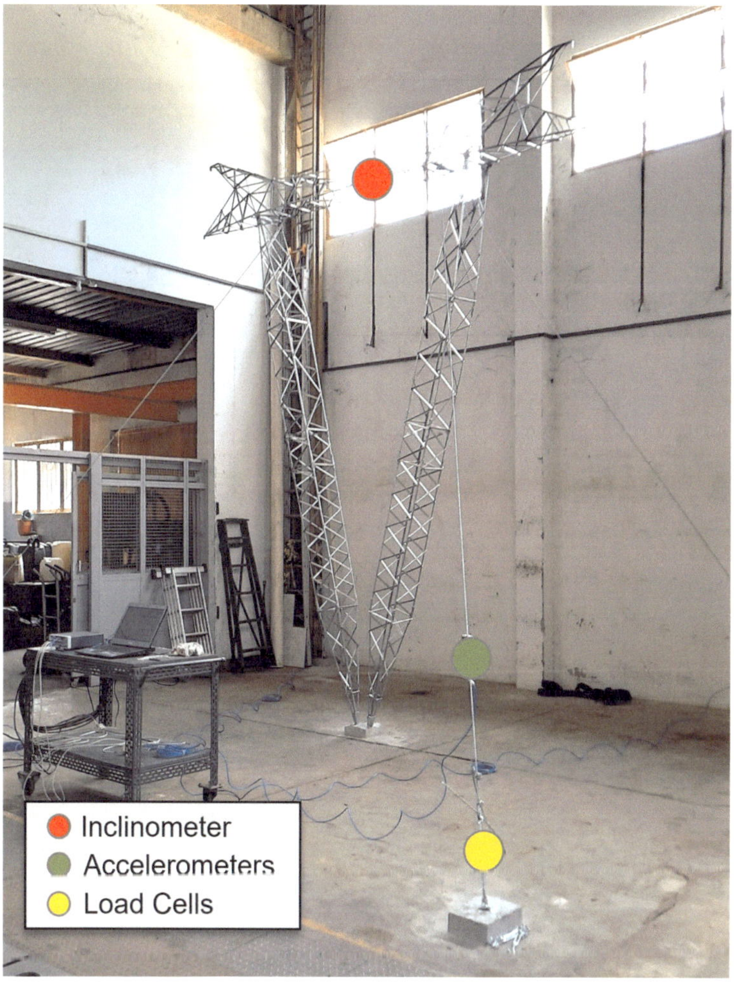

Figure 4.
Structure of the mockup tower assembled in LabEDin, with the indication of the sensor positioning.

Considering small oscillations of the tower, it is possible to define the partial derivatives of the kinetic energy of Eq. (11).

$$\frac{\mathrm{d}}{\mathrm{d}t}\left(\frac{\partial T}{\partial \dot{\theta}_x}\right) = \left(mz_{cg} + I_{xx}\right)\ddot{\theta}_x \quad (12)$$

$$\frac{\mathrm{d}}{\mathrm{d}t}\left(\frac{\partial T}{\partial \dot{\theta}_y}\right) = \left(mz_{cg} + I_{yy}\right)\ddot{\theta}_y \quad (13)$$

$$\frac{\mathrm{d}}{\mathrm{d}t}\left(\frac{\partial T}{\partial \dot{\theta}_z}\right) = I_{zz}\ddot{\theta}_z \quad (14)$$

With the moments of inertia referring to the center of gravity of the model, the total potential energy can be defined according to the Eq. (15).

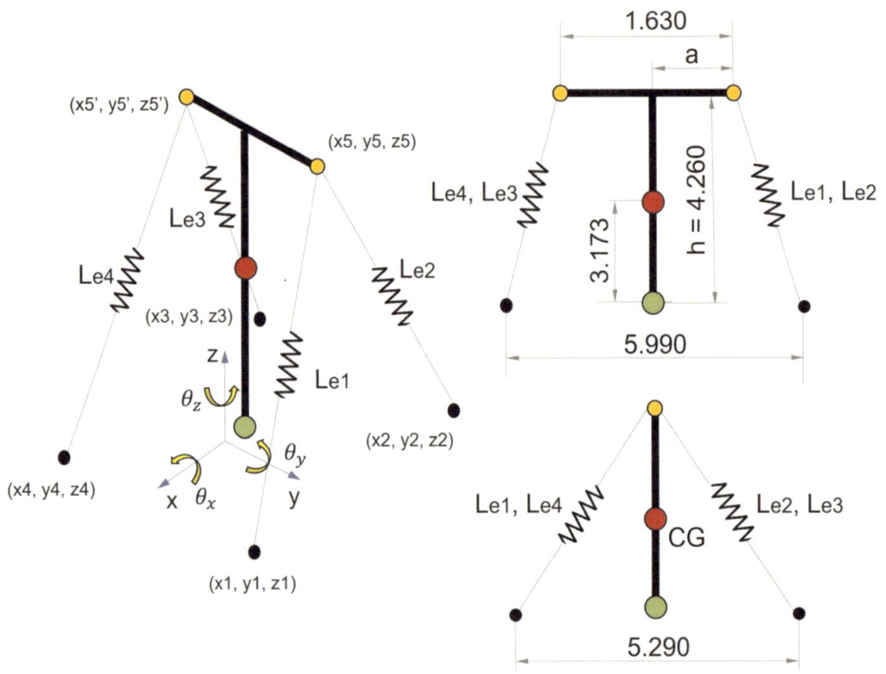

Figure 5.
Simplified structure considered to build the dynamic model of the tower for the EKF application.

$$U_g + U_K + U_{K_t} = m\ g\ z_{cg} \cos\theta_x \cos\theta_y + \sum_{j=1}^{4} \frac{1}{2} k_j (L_c - L_{cn})_j^2 + \cdots$$
$$\cdots + \frac{1}{2} k_{t_x} \theta_x^2 + \frac{1}{2} k_{t_y} \theta_y^2 + \frac{1}{2} k_{t_z} \theta_z^2 \tag{15}$$

In our case, it is possible to consider only small oscillations of the tower and disregard the torsional stiffness around the base pivot point, without loss of accuracy in the results. Thus, the total potential energy is equivalent to the elastic potential energy and its partial derivatives are shown in Eq. (16):

$$\frac{\partial U}{\partial \theta_{x,y,z}} = \frac{\partial U_K}{\partial \theta_{x,y,z}} = \sum_{j=1}^{4} k_j (L_c - L_{cn})_j \frac{\partial L_{c_j}}{\partial \theta_{x,y,z}} \tag{16}$$

Finally, based on the definition of Eq. (9), it is possible to determine the system of second order differential equations for the tower motion according to Eqs. (17)-(19):

$$\left(m\ z_{cg} + I_{xx}\right)\ddot{\theta}_x - \sum_{j=1}^{4} k_j (L_c - L_{cn})_j \frac{\partial L_{c_j}}{\partial \theta_x} = Q_{\theta_x} \tag{17}$$

$$\left(m\ z_{cg} + I_{yy}\right)\ddot{\theta}_y - \sum_{j=1}^{4} k_j (L_c - L_{cn})_j \frac{\partial L_{c_j}}{\partial \theta_y} = Q_{\theta_y} \tag{18}$$

$$I_{zz}\ddot{\theta}_z - \sum_{j=1}^{4} k_j (L_c - L_{cn})_j \frac{\partial L_{c_j}}{\partial \theta_z} = Q_{\theta_z} \tag{19}$$

The numerical values of the tower properties are: $m = 90.7 kg$, $I_{xx} = 269.53 kg\, m^2$, $I_{yy} = 194.95 kg\, m^2$, $I_{zz} = 76.53 kg\, m^2$. Considering the free vibrations of the tower, the dynamic equations of the movement can be written as shown in Eq. (20):

$$\begin{cases} M_x\ddot{\theta}_x = -C_x\dot{\theta}_x - K_x f(\theta_x, \ \theta_y, \ \theta_z) \\ M_y\ddot{\theta}_y = -C_y\dot{\theta}_y - K_y f(\theta_x, \ \theta_y, \ \theta_z) \\ M_z\ddot{\theta}_z = -C_z\dot{\theta}_z - K_z f(\theta_x, \ \theta_y, \ \theta_z) \end{cases} \tag{20}$$

It is important to highlight that the values in $\boldsymbol{M}$, $\boldsymbol{K}$ and $\boldsymbol{C}$ are dependent on the variations in all DOF of the corresponding axes $(\boldsymbol{\theta}_{\mathbf{x}},\boldsymbol{\theta}_{\mathbf{y}},\boldsymbol{\theta}_{\mathbf{z}})$. The Eq. (21) expresses the dynamic behavior of tower, through the first order differential state equations for free vibrations.

$$\dot{x}(t) = Ax(t) \tag{21}$$

With:

$$x(t) = \begin{bmatrix} \theta_x & \theta_y & \theta_z & \dot{\theta}_x & \dot{\theta}_y & \dot{\theta}_z \end{bmatrix}^T \tag{22}$$

The time-variant state transition matrix $\boldsymbol{A}$ has the following aspect:

$$A = \begin{bmatrix} 0 & I \\ -M^{-1}K & -M^{-1}C \end{bmatrix} = \begin{bmatrix} A_{(1,\,1)} & A_{(1,\,2)} \\ A_{(2,\,1)} & A_{(2,\,2)} \end{bmatrix} \tag{23}$$

And the sub-matrices $\boldsymbol{A}_{(2,\,1)}$ and $\boldsymbol{A}_{(2,\,2)}$ can be defined respectively according to Eqs. (24) and (25).

$$A_{(2,\,1)} = \begin{Bmatrix} -\dfrac{K_{1,1}}{M_x} & -\dfrac{K_{1,2}}{M_x} & -\dfrac{K_{1,3}}{M_x} \\ -\dfrac{K_{2,1}}{M_y} & -\dfrac{K_{2,2}}{M_y} & -\dfrac{K_{2,3}}{M_y} \\ -\dfrac{K_{3,1}}{M_z} & -\dfrac{K_{3,2}}{M_z} & -\dfrac{K_{3,3}}{M_z} \end{Bmatrix} \tag{24}$$

Where $\boldsymbol{K}_{i,j}$ depends on the cable characteristics and disposition.

$$A_{(2,\,2)} = \begin{Bmatrix} -\dfrac{C_x}{M_x} & 0 & 0 \\ 0 & -\dfrac{C_y}{M_y} & 0 \\ 0 & 0 & -\dfrac{C_z}{M_z} \end{Bmatrix} \tag{25}$$

5. Tower movements estimate based on EKF algorithm

During 1 year, several tests were carried out in the UNICAMP's laboratory to simulate the foundation failures of the guyed tower. Systematic relaxation of the stay cables was introduced by unscrewing the tensioners close to each base. **Figure 6** presents the cable assembly and the position of the devices. The measurements of the accelerometers, load cells and inclinometer are made in time-domain every 15 minutes. **Table 1** shows the specification of the sensors used in the tower monitoring.

The signal acquisition of all sensors was performed using LMS SCADAS Mobile SCR05® manufactured by Siemens, which has a total of 40 ADC channels. The system was configured with a frequency sampling rate of 256 Hz and an acquisition time of 64 seconds for each measurement. Every sensor generates signals with 16,384 points per channel and **Figure 7** presents the picture of the data acquisition system.

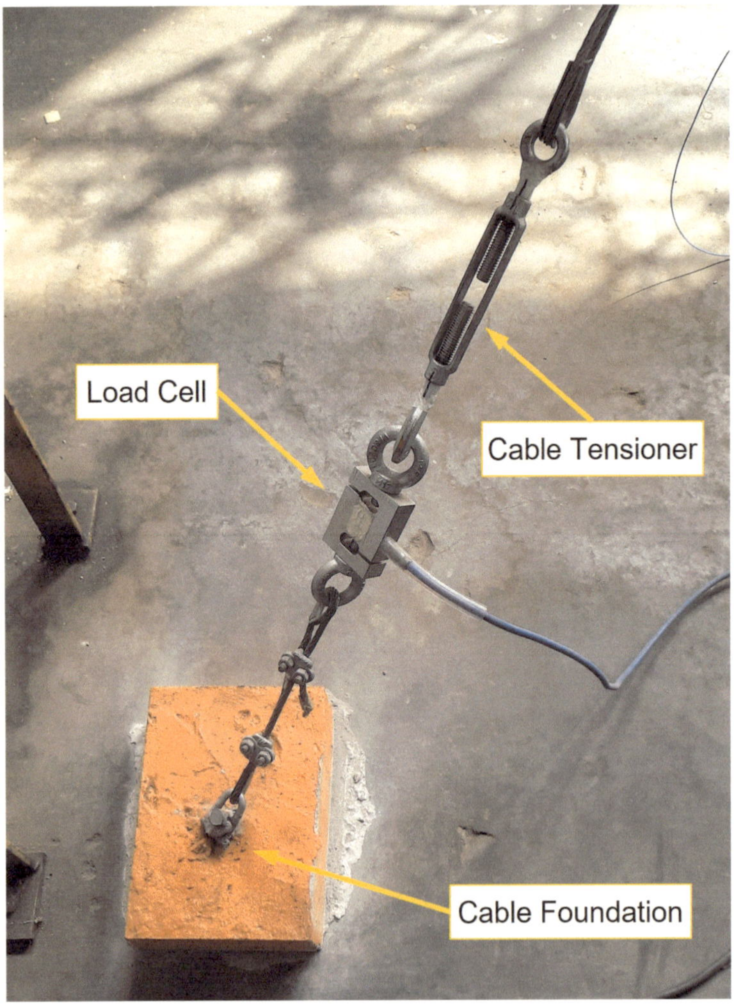

Figure 6.
Devices for the force measurements and load application at the stay cable.

Sensor	Model/Characteristics
Accelerometer[a]	PCB Piezotronics® 356A15, ICP®, Triaxial
Load cell[b]	Alfa Instruments®, S-type, Stainless Steel, Capacity of 500 kg
Inclinometer[a]	Fredericks Company®, 0729-1760-04, Interface RS485 USB, Dual-axis

[a]EKF estimations consider the sensor fusion from the signals of the accelerometers and inclinometer.[b]Load cells are used only to validate the forces calculated indirectly by the accelerometers.

Table 1.
Specification of the sensors used for the measurements.

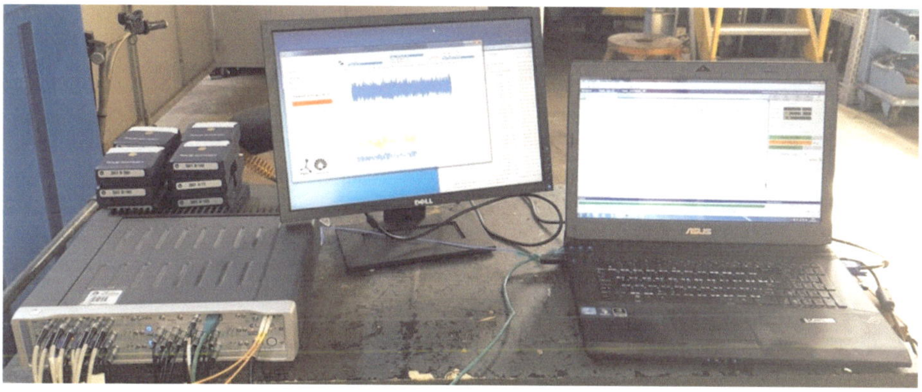

Figure 7.
Simens LMS SCADAS® used for the signal acquisition of the mockup tower.

5.1 Definition of the extended Kalman filter parameters

The variation of the stay cable force promote strain variations, which are converted into displacements projected on plane (X, Y) at the connecting points of two cables to the tower, i.e. Points 5 and 5′ presented in **Figures 5** and **8**. The Eqs. (26) and (27) present the formula to calculate the displacements based on the cable force determined by the measured natural frequency.

$$\delta_5 = \left[\left(\frac{F_2}{k_2} + L_{cn_2}\right) - L_{f_2} - \left(\frac{F_1}{k_1} + L_{cn_1}\right) + L_{f_1}\right] \div 2 \tag{26}$$

$$\delta_{5'} = \left[\left(\frac{F_3}{k_3} + L_{cn_3}\right) - L_{f_3} - \left(\frac{F_4}{k_4} + L_{cn_4}\right) + L_{f_4}\right] \div 2 \tag{27}$$

Where: F_j = Force at cable j.

The stiffness of the cable presented in Eqs. (26) and (27) can be calculated according to Eq. (28) and considering the characteristics of the steel wire rope used in the mockup tower (Φ=14″, class 6×7, fiber core), the equivalent elasticity modulus and the cross section area are respectively 88,200 MPa and 16.1792 mm^2. This information can be verified in [8].

$$k_j = \frac{E_c A_c}{L_{f_j}} \tag{28}$$

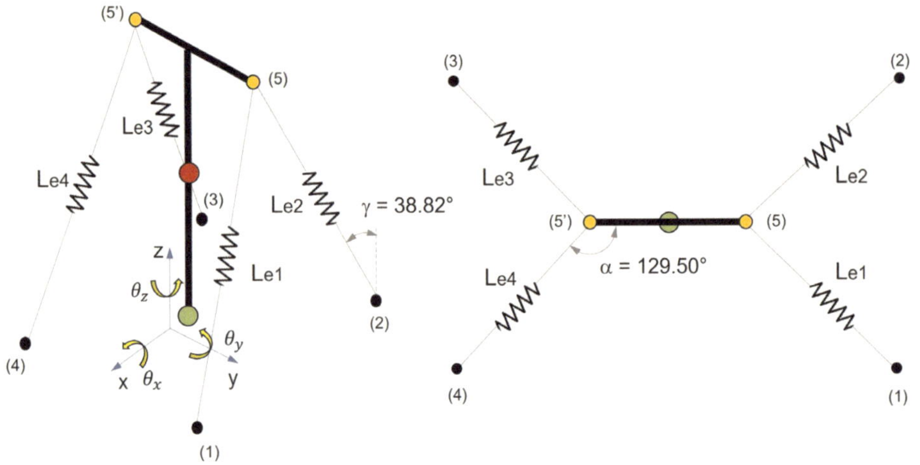

Figure 8.
Dynamic model of the tower. Top view details.

With the displacements of these points it is possible to define the rotation angles of the structure in respect to X,Y axes. The displacements not related to variations in the average cable force, e.g. moderate wind condition, can be quantified according to Eqs. (29)-(31). These angles correspond to the indirect measured elements of vector $\boldsymbol{z}_\delta$, considering $\bar{\boldsymbol{\delta}}_5$ as the average displacement of Points 5 and 5′.

$$z_{\delta_x} = \arcsin\left[\frac{\delta_5 \cos(90-\gamma)\cos(180-\alpha)}{h}\right] \tag{29}$$

$$z_{\delta_y} = \arcsin\left[\frac{\delta_5 \cos(90-\gamma)\sin(180-\alpha)}{h}\right] \tag{30}$$

$$z_{\delta_z} = \arcsin\left[\frac{\delta_5 \cos(90-\gamma)\sin(180-\alpha)}{a}\right] \tag{31}$$

On the other hand, the stay cable relaxation promotes a decrease in average force $\overline{\boldsymbol{F}}$ and this reduction is also responsible to generate proportional rotations of the tower in respect to X,Y axes. From the mathematical model of the tower we can define the Eqs. (32) and (33), which present these relations. It is possible to observe, that the equations indicate zero inclinations when the average force is approximately 2600 N, which is the assembly preload of the cables.

$$z_{F_x} = 4.32 \cdot 10^{-3} - 1.659 \cdot 10^{-6}\overline{F} \tag{32}$$

$$z_{F_y} = 4.91 \cdot 10^{-3} - 1.883 \cdot 10^{-6}\overline{F} \tag{33}$$

The sensor fusion technique is shown in **Figure 9**. It presents the input measurements consisted by the inclinometer angles $\boldsymbol{z}_{\bar{\theta}_x}, \boldsymbol{z}_{\bar{\theta}_y}$ and the stay cable forces $\overline{\boldsymbol{F}}$, which are calculated from the natural frequencies $\boldsymbol{\omega}_\mathbf{n}$ determined by the accelerometer measurements.

A question that often comes up is why not just measure angles using the inclinometer to monitor the structural health of the tower. The reason is that there may be a situation where two symmetrically opposite guyed cables exhibit approximately the

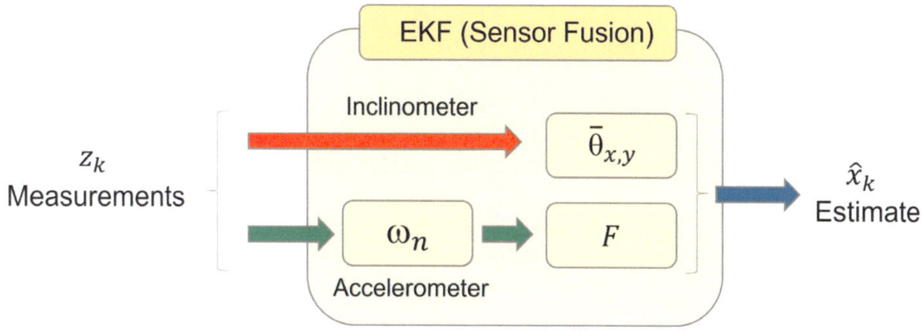

Figure 9.
Sensor fusion considered to estimate the tower movements.

same amount of relaxation. In this case, the position of the top center of the tower presents practically no angles measured by the inclinometer in X,Y directions, but the average stay cable force will decrease. This situation can lead to an unstable structural condition with high-intensity winds affecting the tower. Thus, only inclination angles of the guyed tower cannot inform the actual integrity of its structure.

According to the presented methodology, the extended Kalman filter can only estimate the absolute value of the displacement. Fusing this information with the inclinometer data, it is possible to define the direction of this displacement and it also helps to indicate the cable that presented the failure. The tower inclinations due to the thermal expansion of the structure can be verified by the consideration of the inclinometer measurements.

The state transition matrix changes with rotation angles $\boldsymbol{\theta}_{\mathbf{x}}$, $\boldsymbol{\theta}_{\mathbf{y}}$ and $\boldsymbol{\theta}_{\mathbf{z}}$, which are estimated for each interaction of the algorithm after the calculation of the cable forces from the measured natural frequencies. In our case, the estimate vector is defined according to:

$$\hat{x}_k = \begin{bmatrix} \hat{\theta}_x & \hat{\theta}_y & \hat{\theta}_z & \hat{\dot{\theta}}_x & \hat{\dot{\theta}}_y & \hat{\dot{\theta}}_z \end{bmatrix}^T \quad (34)$$

In the sequence, the matrices considered in the EKF algorithm will be presented. These matrices are determined from known analytical results of the tower dynamic model, i.e., by applying forces at the tower structure and considering its responses as targets for EKF estimates. The state matrix for the measurements:

$$H = \begin{bmatrix} 0.50 & 0 & 0 & 0 & 0 & 0 \\ 0 & 0.47 & 0 & 0 & 0 & 0 \\ 0 & 0 & 1 & 0 & 0 & 0 \end{bmatrix} \quad (35)$$

The noise covariance matrix of the state transition:

$$Q = 10^{-3} \cdot \begin{bmatrix} 1 & 0 & 0 & 0 & 0 & 0 \\ 0 & 1 & 0 & 0 & 0 & 0 \\ 0 & 0 & 1 & 0 & 0 & 0 \\ 0 & 0 & 0 & 1 & 0 & 0 \\ 0 & 0 & 0 & 0 & 1 & 0 \\ 0 & 0 & 0 & 0 & 0 & 1 \end{bmatrix} \quad (36)$$

The covariance matrix of the measurement noise:

$$R = 5 \cdot \begin{bmatrix} 1 & 0 & 0 \\ 0 & 1 & 0 \\ 0 & 0 & 1 \end{bmatrix} \quad (37)$$

Note: Null values were adopted as initial conditions for EKF estimate and error covariance.

The Eq. (38) shows the measurement vector at instant k considered as an input to the EKF, considering the weighting factors $\Psi_\delta, \Psi_F, \Psi_{\overline{\theta}}$ for each sensor quantity.

$$z_k = \Psi_\delta \begin{bmatrix} z_{\delta_x} \\ z_{\delta_y} \\ z_{\delta_z} \end{bmatrix} + \Psi_F \begin{bmatrix} z_{F_x} \\ z_{F_y} \\ 0 \end{bmatrix} + \Psi_{\overline{\theta}} \begin{bmatrix} z_{\overline{\theta}_x} \\ z_{\overline{\theta}_y} \\ 0 \end{bmatrix} \quad (38)$$

In our case, the given weights to the cable force data are $\Psi_\delta = 0.48$; $\Psi_F = 0.10$ and the weight for the inclinometer measurements is $\Psi_{\overline{\theta}} = 0.42$.

5.2 Experimental procedure

Now will be presented some simulation of failures, which can occur in the real tower due to the degradation of the foundation, caused as an example, by the erosion of the soil. The tensioners of two symmetrically opposite cables were unscrewed, promoting the relaxation of the stays. Each turn of the tensioner generates a displacement of 4.20 mm along the direction of the cable.

Initially, the stay cables are fixed with an average force of 2600 N with the inclinometer indicating zero degree with respect to X and Y axes. **Figure 10** presents

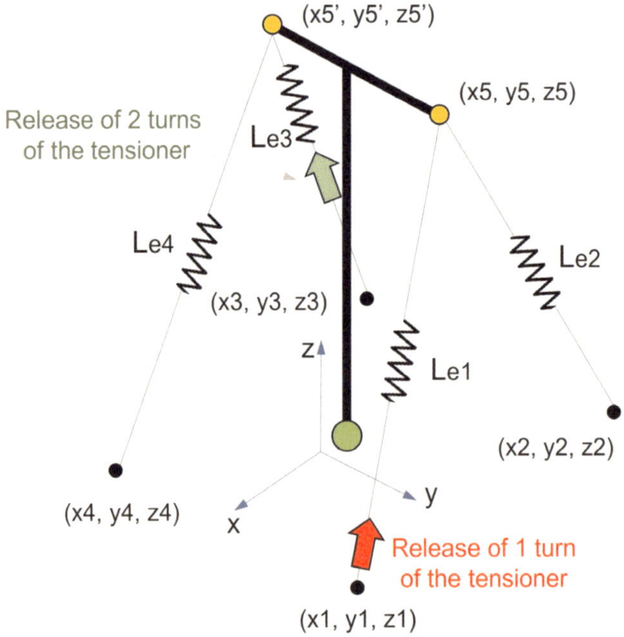

Figure 10.
Locations of the interventions to simulate the failures.

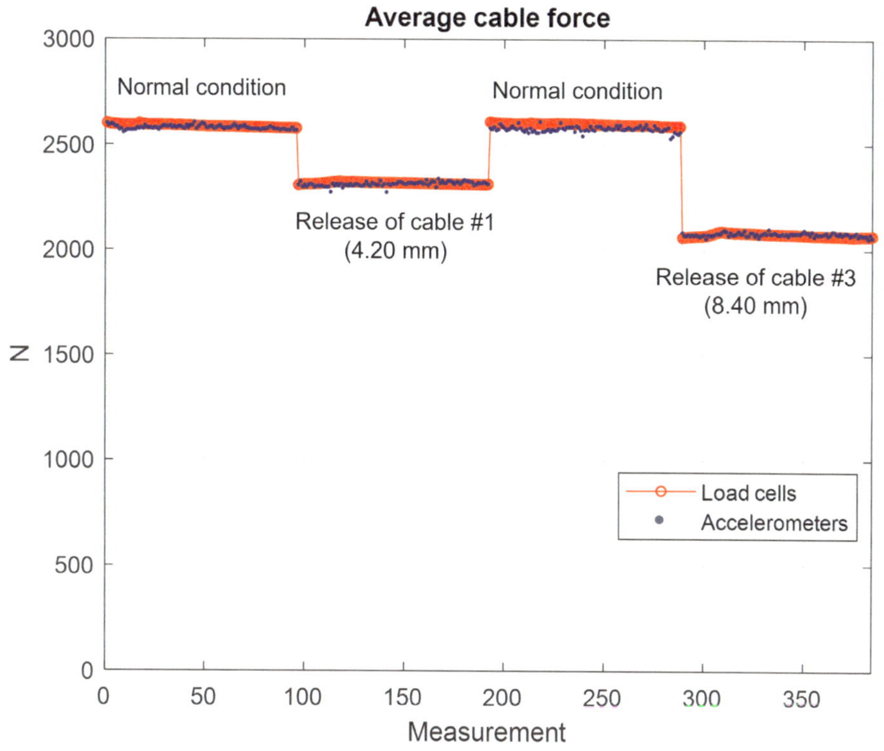

Figure 11.
Comparison between average cable forces: measured with the load cells and calculated with Eq. (8).

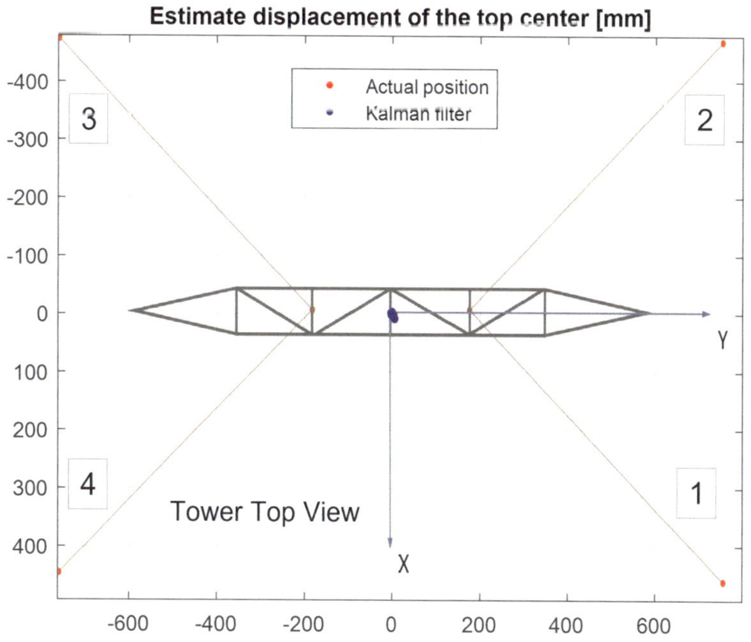

Figure 12.
Tower top view showing the stay cable numbering and the considered coordinate system. In the central region there are the estimate displacements.

the sequence of events to estimate the position of the top center of the tower. It consists of 96 measurements performed during every 24 h, when some manual intervention was made on the cable tensioners.

In this case, after the relaxation of Cable 1, the EKF estimates the position of the top center of the tower. Subsequently, the tensioner was re-tightened to its initial position and then, the tensioner of Cable 3 was released for new estimates of the tower position. In a real situation, this test represents two failures separated by one tower maintenance intervention.

After 4 days, the test was completed and **Figure 11** compares the average force on the stay cables measured with the load cells, with the average force calculated from the natural frequencies obtained from the accelerometer measurements.

A top view of the tower showing the numbering of the support cables, the considered coordinate system and the estimated displacements in the central region based on Extended Kalman filter can be seen in the **Figure 12**.

A detailed view of the estimated displacements for each different condition imposed to the mockup tower can be seen in **Figure 13**. It is possible to observe three groups of points. The central group of points (in a red ellipse) shows the top position for the stay cables in normal conditions, i.e. immediately after assembly with an average force of approx. 2600 N. On the left, it is possible to see a group of points inside of a dashed green ellipse, which presents the positions after the release of the

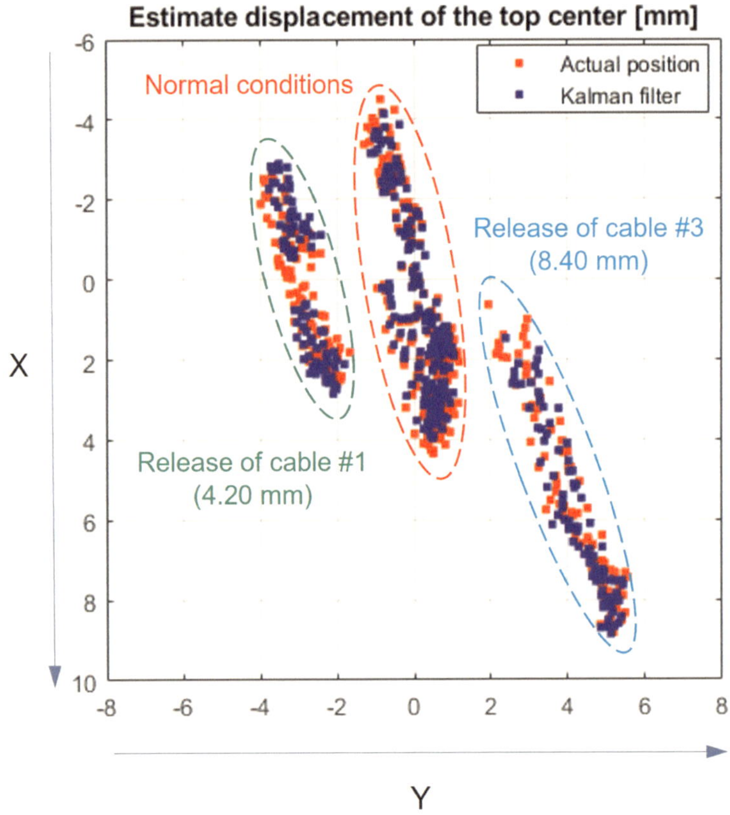

Figure 13.
Detail of the estimate displacements of the tower top center based on the extended Kalman filter algorithm.

Cable 1 tensioner by one turn. As expected for this case, the top center moves slightly to the location symmetrically opposite of the foundation of Cable 1.

Starting from the second normal condition of the tower (after re-tightening the tensioner of Cable 1), there was a release of two turns in the tensioner of Cable 3, that moves slightly the top center position to the symmetrically opposite location of the Cable 3 foundation and these points are shown within a dashed cyan ellipse on the right side of the figure.

Continuing the analysis of **Figure 13**, it is also possible to observe the variations of the top position within each group of points, mainly along X-axis direction. These variations are due to the fluctuations in lab temperature during the 24 hours of each test. Basically, when the temperature increases the tower bends in the +X direction, while at night with the decrease in temperature the tower bends in the -X direction. The non-perfect symmetry of the mockup tower and the oblique incidence of sun rays during the day, contribute to the lateral movements in the Y axis direction.

Figure 14 summarizes the adopted procedure, which starts by considering the accelerometer measurements in time-domain and ends with the EKF estimate for the tower top position. It is important to note that the code is able to evaluate the structural health of the guyed tower, not only for large movements of the top, but also for low average force on the stay cables.

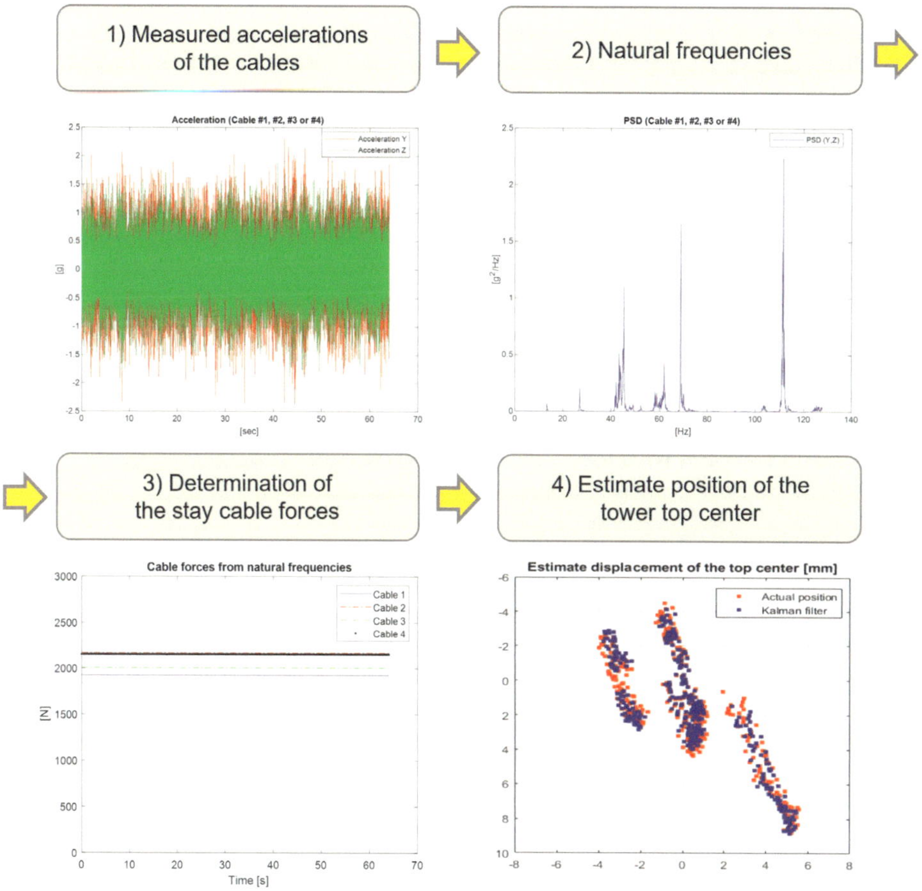

Figure 14.
Procedure to estimate of the tower top center position starting from cable acceleration measurements.

6. Conclusions

The EKF was considered to estimate the displacements of the upper part of a mockup of a guyed tower in the laboratory. A MATLAB code was developed to execute the algorithm that indicates the behavior of the structure with a very good accuracy.

Stay cable relaxation is the type of failure that is judged to be more usual in the field. This situation was simulated extensively with good approximation between real and estimate values. The equation defined to establish the relation between the natural frequency and the cable force has indicated to be a reliable procedure, as an alternative substituting load cells to evaluate variations of the cable force over the time. The load cell is recommended only for initial measurements of the stay cable preload, for the adjustment of the cable equivalent length. After this step, the cable force can be monitored only by the accelerometers.

The influence of the transmission line conductors in the dynamic responses of the structure of the tower is considered negligible and due to this fact, the mockup tower in the laboratory does not have elements that simulate the influence of the conductors. According to [9], the electrical insulators promote some decoupling of the eolian vibrations that excite the transmission lines.

Sensor fusion considering the accelerometers and inclinometer, proved to be the best option to indicate the structural integrity of the mockup tower considering the stay cable relaxation, responses to external loads and tower thermal expansion. The individual consideration of those sensors cannot provide a reliable report about the guyed tower health. As explained previously, the inclinometer alone cannot indicate the inclination of the structure if two symmetrically opposite cables are losing preload. From the obtained results, the behavior of the real guyed towers is expected to be very similar to those simulated in the laboratory.

Acknowledgments

The authors are grateful to R&D Project PD-07130-0047 "Evaluation of Multivariable Modeling Methods for Monitoring the Health of Guyed Towers in Overhead Power Lines", partially funded by: Transmissora Aliança de Energia Elétrica S.A. (TAESA), with resources from ANEEL R&D Program.

Conflict of interest

The authors declare no conflict of interest.

Nomenclature

$\hat{x}_k$	estimated value at instant k
$P_{\bar{k}}$	error covariance at instant k
A	state transition matrix
Q	noise covariance matrix of the state transition
K_k	Kalman gain at instant k
R	covariance matrix of the measurement noise

H	state matrix for the measurement
z_k	measured signal at instant k
z_δ	rotation signals from the cable strain
z_F	rotation signals from average the cable force
$z_{\overline{\theta}}$	rotation signals from the inclinometer
ω_n	natural frequency
n	mode of vibration
L_e	equivalent cable length
F	cable tensile force
μ	cable linear mass density
T_v	translational kinetic energy
T_ω	rotational kinetic energy
U_g	gravitational potential energy
U_K	linear elastic potential energy
U_{K_t}	torsional elastic potential energy
Q	external load
v_{cg}	velocity of the tower's center of gravity
z_{cg}	height of the tower's center of gravity
m	mass of the tower structure
g	gravitational acceleration
$I_{xx,yy,zz}$	moments of inertia referred to the center of gravity
θ	tower angles (in x,y,z axes)
M	mass matrix
C	damping matrix
K	stiffness matrix
k	stay cable stiffness
k_t	tower torsional stiffness at pivot base point
h	height of the tower
a	half length of the tower upper structure
L_c	stay cable instantaneous length
L_{cn}	stay cable assembly length with preload
L_f	stay cable free length
E_c	equivalent Young modulus of the cable
A_c	equivalent cross sectional area of the cable
$\Psi_{\delta,F,\overline{\theta}}$	weighting factors for the measured quantities

Abbreviations

DOF	degrees of freedom
EKF	extended Kalman filter
IoT	Internet of Things
LabEDin	dynamic testing laboratory
PSD	power spectrum density
RF	radio frequency
UNICAMP	University of Campinas

Author details

Alexandre Schalch Mendes[1*†], Pablo Siqueira Meirelles[1†], Janito Vaqueiro Ferreira[1†] and Eduardo Rodrigues de Lima[1,2†]

1 University of Campinas, Campinas, Brazil

2 Instituto de Pesquisas Eldorado, Campinas, Brazil

*Address all correspondence to: schalch@unicamp.br

† These authors contributed equally.

References

[1] Särkkä S. Bayesian Filtering and Smoothing. UK: Cambridge University Press; 2013. DOI: 10.1017/CBO9781139344203. ISBN: 978-1107 619289

[2] Kim P. Kalman Filter for Beginners: With MATLAB Examples. Scotts Valley, CA, USA: CreateSpace Independent Publishing Platform; 2011; ISBN: 978-1463648350

[3] Kim Y, Bang H. Introduction to Kalman filter and its applications. In: Govaers F, editor. Introduction and Implementations of the Kalman Filter. London: IntechOpen; 2018. DOI: 10.5772/intechopen.80600

[4] Julier S, Uhlmann JK, Durrant-Whyte H. A new approach for filtering nonlinear systems. Proceedings of 1995 American Control Conference - ACC'95. Vol. 3. 1995. pp. 1628-1632. DOI: 10.1109/ACC.1995.529783

[5] Santos AS. Estudo de Vibrações Eólicas em Linhas de Transmissão de Energia Elétrica de Alta Tensão [thesis] (in Portuguese). Federal University of Pará, Brazil. 2008

[6] Mendes AS, Ferreira JV, Meirelles PS, Nóbrega EGO, Lima ER, Almeida LM. Evaluation of multivariable modeling methods for monitoring the health of guyed towers in overhead power lines. Sensors. 2021;**21**(18):6144. DOI: 10.3390/s21186144

[7] Doughty S. Mechanics of Machines. NY, USA: John Wiley and Sons Inc.; 1988. p. 467. ISBN: 0-471-84276-1

[8] CIMAF, Manual Técnico de Cabos (in Portuguese). Belgo Bekaert Arames, ArcelorMittal [Internet]. 2012. Available from: https://www.aecweb.com.br/cls/catalogos/aricabos/CatalogoCIMAF2014Completo.pdf. [Accessed: 2022-June-07]

[9] Mathur RK, Shah AH, Trainor PG, Popplewell N. Dynamics of a guyed transmission tower system. IEEE Transactions on Power Delivery. 1987; **2**(3):908-916. DOI: 10.1109/TPWRD.1987.4308195

Chapter 3

Detection and Localization of a Failure in a Pipeline Using a Kalman Filter: An Intelligent Integrated Approach Powered by Bayesian Classification

Rajamani Doraiswami and Lahouari Cheded

Abstract

An integrated approach, based on the fusion of Model-Based Approach (MBA) and Model-Free Approaches (MFA) and powered by Bayesian classification, is proposed to ensure high probability of correct estimation of leakage detection and localization with low false alarm probability to prevent disastrous consequences to the economy and environment. To ensure mathematical tractability, the nonlinear model is better approximated using linear parameter-varying (LPV) model at various operating points indicated by scheduling variables. Flows at various pipeline sections are measured and transmitted wirelessly to a monitoring station. If there is a difference in the flows across a section, it indicates a leakage, and a drone is then sent to determine the exact location of the leakage. The pipeline trajectory is accurately estimated by a human operator. Using the input and the trajectory output, termed signal, an Autonomous Kalman filter (AKF) is designed to ensure accurate tracking of the desired trajectory. The emulator-generated data are used to identify the system, complement historical data to MFA, and develop the classifier fusion. The leakage is sequentially diagnosed by judiciously selecting the most appropriate approach (MFA or MBA) to ensure a fast and accurate diagnosis. The proposed scheme was evaluated on a physical system.

Keywords: leakage diagnosis, emulators, emulator generated data, Kalman filter, sequential diagnosis, Bayes' classifier fusion, trajectory tracking, nonlinear two-tank model, linear parameter-varying model, signal model, disturbance model, measurement noise, model-based approach, model-free approach

1. Introduction

The pipelines are widely used for transporting fluids such as water or petroleum products such as fossil fuels, gases, chemicals, and other essential hydrocarbon fluids. The pipeline network covers thousands of kilometers. The effect of leak manifests as a sudden decrease in the pressure in the flow rate of fluid being transferred. Leakage in pipes and storage tanks occurs due to factors such as faulty joints, excessive stress, aging, and holes caused by corrosions [1–10]. The leakage detection methods consist of manual inspection by trained linesmen, satellite imaging, and in recent years by autonomously guided drones flying over the pipeline route. The drone performs pipeline condition assessment and mechanical damage and cracking on above ground structures. It can be designed to detect fatigue cracking, corrosion, or other defects that cannot be observed from ground.

Classifier fusion: A Model-Based Approach (MBA) and different Model-Free schemes, lumped under the heading of Model-Free Approach (MFA) are used here. The MBA includes a Kalman filter (KF), an extended Kalman filter (EKF), an observer, and a system identification stage. For a process with an unknown model, the model-free approaches are used. In practical situations, a fusion of model-based and model-free approach; combination of the analytical and knowledge-based methods may be the most appropriate solution.

Physical system: A wider class of physical dynamic systems is nonlinear containing nonlinearities such as saturations, rate limiters, dead-zones, backlash, and turbulence. The analysis, design, estimation, identification, and control of nonlinear system are not mathematical intractable. As there is a wealth of tools for the analysis and design of linear systems, in the recent years, the Linear Parameter-Varying (LPV) systems have received a lot of attention [11, 12]. The piecewise-linear model approach helps develop computationally simple, efficient, and robust schemes for identification, design of Kalman filter, fault detection, and isolation. The LPV paradigm has become a standard formalism in systems and control, for analysis, synthesis of controllers, and even system identification.

The output of the system is a sum of signal, disturbance, and measurement noise. A signal is the desired waveform while the disturbance and the measurement noise are termed as "noise." Wind gusts, pressure variations, and fluctuations in the flow affect the system output and are all treated as system disturbances whose effects are to be mitigated at least [2–8]. It is assumed that the stochastic disturbance and measurement noise are zero mean Gaussian processes, and that the signal, disturbance, and measurement noise are mutually uncorrelated with each other.

The principle states output will track desired trajectory *if and only* if the structure of a controller contains a) an internal model of the desired trajectory driven by the tracking error between the output and desired trajectory, and b) the closed loop system formed of the plant and the internal model is asymptotically stable.

The internal model principle governs the structure of the Kalman filter, which state that the residual is a zero mean white noise process *if and only if* the Kalman filter is a copy of the system model and driven by the residual, which the error between the output and its KF estimate. The optimality and robustness of the KF estimates are two important features of our proposed integrated approach which are both discussed in detail in [13].

1.1 Kalman filter and its properties

The KF forms the backbone of the proposed detection and localization scheme in view of its key properties [12–18].

Internal model structure: The principle states output will track desired trajectory if and only if the structure of a controller contains a) an internal model of the desired trajectory driven by the tracking error between the output and desired trajectory, and

b) the closed loop system formed of the plant and the internal model is asymptotically stable.

The internal model principle governs the structure of the Kalman filter, which states that the residual is a zero mean white noise process *if and only* if the Kalman filter is a copy of the system model and driven by the residual, which the error between the output and its KF estimate. The optimality and robustness of the KF estimates are two important features of our proposed integrated approach, which are both discussed in detail in [13].

1.2 Identification using the residual of the KF

The fundamental requirement of identification is that the leftover from identification, namely the residual, is a zero-mean white noise process that contains no information. To meet this requirement, the following model-matching property of the Kalman filter

- The identified model is accurate *if and only if* the KF residual is a zero mean white noise process

Hence, the residual of model of the KF associated with the system is identified. The order of the identified model is determined from the minimal order that ensures that the identification error is a zero mean white noise process. Further there is no need for an a priori knowledge of the variance of the disturbance and measurement noise avoiding thereby solution of Riccati equation.

In model-based, model-free, and classifier fusion approach, it is crucial to provide representative and sufficient data covering the normal, perturbed, and fault-bearing operating scenarios resulting from the variations of the subsystems. As the parameters' subsystems are generally inaccessible, the data are generated indirectly by performing several off-line experiments to mimic likely operating scenarios. In model-based scheme, the emulator-generated data are used in identification of the system and the associated Kalman filter to ensure that identified models are robust to model perturbation and are significantly more accurate compared with that obtained using the classical approach of using merely the input and the output without including the perturbed models [13, 14].

The model-free approaches include Limit Checking, Visual, and Plausibility (LVP) analysis, Artificial Neural Network (ANN), Fuzzy Logic (FL), and Adaptive Neuro-Fuzzy Inference System. (ANFIS) [13–15]. The model-based approach using Kalman filtering is widely used for fault diagnosis [4, 6, 10, 12–18]. The model-free approach can readily learn the distinguishing features that help classify the system as either normal or abnormal and then isolate the faulty subsystem. However, these model-free approaches suffer from some disadvantages. For ANN, there is a lack of transparency, a need for a long-, and rich-enough training data covering most, if not all, operational scenarios, and a possibly lengthy training time. Although more transparent than ANNs, FL techniques face the problem of expressing the knowledge in the form of "if-and then" rules from the vast amount of data and from the experts' knowledge and experience. For more accuracy, the number of these rules can increase to an unacceptably large number. To overcome this problem, a combination of ANN and FL, termed ANFIS, has been proposed in recent years. However, the problem of detecting incipient faults, their fault size, and predicting the occurrence of a fault using these model-free approaches remains an unsolved and important challenge. On the other hand, the model-based method could detect and isolate incipient faults as the model captures the complete behavior of the static and the dynamic behaviors of the system. The model predicts the behavior of the system as well as unforeseen operating scenarios, including total failure, with good accuracy. However, the behavior of a physical system at all operating points, especially the nonlinear ones, cannot be captured accurately in the form of a mathematical model [13].

In model-based, model-free, and classifier fusion approach, it is crucial to provide representative and sufficient data covering the normal, perturbed, and fault-bearing operating scenarios resulting from the variations of the subsystems. As the parameters' subsystems are generally inaccessible, the data are generated indirectly by performing several off-line experiments to mimic likely operating scenarios. In model-based scheme, the emulator-generated data are used in identification of the system and the associated Kalman filter to ensure that identified models are robust to model-perturbation and are significantly more accurate compared with that obtained using the classical approach of using merely the input and the output without including the perturbed models.

The model-free approaches include limit checking, visual, and plausibility analysis (LVP), artificial neural network (ANN), fuzzy logic (FL), and adaptive neuro-fuzzy inference system (ANFIS) [14, 15].

Practical systems are notoriously known to be complex and nonlinear in nature and hence do not lend themselves to mathematically tractable identification, analysis, and design techniques that span the entire operating region. This difficulty is further exacerbated for highly nonlinear systems. This therefore renders the use of the MBA schemes to capture both the static and dynamic behaviors of the system, difficult to use directly on the original system. On the other hand, the MFA schemes, by virtue of their independence of, and hence non-reliance on, a system model, can be readily used to learn the distinguishing features that help classify the system as either normal or abnormal and then isolate the faulty subsystem. However, these model-free approaches suffer from some disadvantages. For ANN, there is a lack of transparency, a need for a long- and rich-enough training data covering most, if not all, operational scenarios, and a possibly lengthy training time. Although more transparent than ANNs,

Fuzzy Logic (FL) techniques face the problem of expressing the knowledge in the form of "if-and then" rules from the vast amount of data and from the experts' knowledge and experience. For more accuracy, the number of these rules can increase to an unacceptably large number. To overcome this problem, a combination of ANN and FL, termed ANFIS, has been proposed in recent years.

However, the problem of detecting incipient faults, their fault size, and predicting the occurrence of a fault using these model-free approaches remains an unsolved and important challenge. On the other hand, the model-based method could detect and isolate incipient faults as the model captures the complete behavior of the static and the dynamic behaviors of the system. The model predicts the behavior of the system as well as unforeseen operating scenarios, including total failure, with good accuracy [10–13]. However, the behavior of a physical system at all operating points, especially the nonlinear ones, cannot be captured accurately in the form of a mathematical model.

The decision of the hypotheses from different classifiers is fused with a view to improving the probability of correct decision with low false alarm probability compared with that obtained by using any one of the classifiers [14]. In the proposed combined approach, the critical information about the presence or absence of a fault is gained in the shortest possible time via the faster model-free schemes such as the LVP.

A more accurate and detailed status of the subsystems is unfolded sequentially by the slower model-based scheme. The combined classifier scheme then fuses the decisions from both approaches, using a Bayes' weighted fusion method [19].

Autonomous Kalman filter: The pipeline is generally laid underground to transmit fluid flow from the source to destination over a very long distance. A trained human operator tracks the trajectory of the pipeline. It is difficult and time-consuming job to track the pipeline trajectory. A Kalman filter is designed using the input and output of the human operator. It is autonomous and replaces human operator and drives the drone accurately, efficiently, and quickly. When a leak is detected in a pipeline section, the drone is sent to the section to detect and locate the exact location from pipeline condition assessment, mechanical damage, and cracking. It can be designed to detect fatigue cracking, corrosion, or other defects that cannot be observed from ground [2].

1.3 Major contributions

The proposed scheme extends the conventional fault diagnosis approach to a wider class of MIMO Box-Jenkins model [13]. As this model is more general than conventional ones, such as AR, MA, and ARMA, it then has wider applications that may include models of systems such as transient flow in pressurized pipes and boiler-steam water flow. The emulator-generated data cover both normal and abnormal operating scenarios including various types of faults.

The emulator-generated data are employed in: (a) the identification of the system, the Kalman filter design, and fault isolation method in the model-based scheme, (b) in training the model-free schemes, and (c) in classifier fusion to ensure that the decisions made by both approaches are based on the same set of sufficient and representative data ensuring that all the diagnosis schemes are provided with a level playing field. In our proposed symbiotic approach, the performance of classifier fusion is significantly superior to that of using only a model-based or a model-free scheme, especially when the system such as a process control system is nonlinear. When the system is operating in the linear region, the performance of the model-based scheme is better while the model-free scheme such as ANN and ANFIS. However, the latter scheme outperforms the former one in the nonlinear operating region. The classifier fusion scheme ensures high probability of correct decision with low false alarm probability. The model-based scheme can detect incipient faults so that a proactive action such as a condition-based maintenance can be taken. Thanks to the availability of a reliable and accurate model, the emulators help predict likely operating fault scenarios.

When a leak is detected in pipeline section, the drone is driven autonomously to that section mimicking the human operator as it were ensuring timely leakage diagnosis, process safety, and environmental protection.

The paper is organized as follows: Section I gives an overview of the proposed scheme covering model-free and model-based schemes. In section 2, the two-tank nonlinear process control system is developed. The system is shown to be governed by a Box-Jenkins model and the identication of associated Kalman filter is developed. Section 3 presents the Kalman filter and its key properties. Section 4 gives details of the sequential fault diagnosis and discusses both the model-free schemes and the classifier fusion. Section 5 gives further details of the model-free schemes. Section 6 discusses only some important details of the model-based schemes as these have been amply discussed in some of referenced previous works. This section also evaluates the successful performance of the proposed scheme on a benchmark laboratory-scale process control system. Finally, section 7 gives the conclusion.

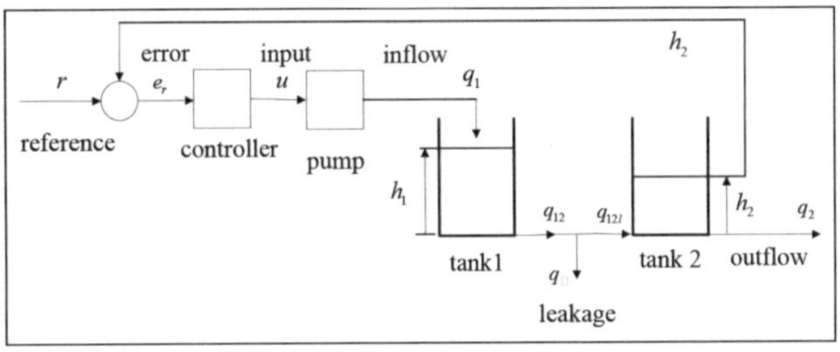

Figure 1.
Two tank process control system.

2. System model

The nonlinear model of the two-tank process control system exhibiting turbulence is approximated by piecewise linear dynamic model at each operating point using LPV approach [20]. **Figure 1** shows a two-tank process control system formed of two tanks, namely process, consumer tank denoted tank 1 and tank 2, respectively, a controller, and a pump. The controller is designed to maintain fluid level h_2(t) at specified reference level r(t) and is driven by the tracking error, e_r(t)=r(t) -h_2(t). The control input u(t) drives a pump to supply the fluid to the tank 1 and $q_1$is the inflow. The fluid level of the tank 1 is h_1(t). A long pipeline connects the two tanks and is r(t) subjected to a leak q_l at some section of the pipe. The outflow q_{12} of the tank 1 and q_{12l} is the inflow to the tank 2.

The tanks are cylindrical, and the height of the process tank h_1 is higher than that of the consumer tank h_2, that is $h_1 \geq h_2$. The two tanks are connected by a long pipeline. The pressures exerted by the tank 1 and 2 at their end of the pipe are respectively $\rho g h_1$ and $\rho g h_2$, where ρ is the density of the fluid and g is the acceleration due to gravity. Since the flow is proportional to the pressure difference, fluid flows from tank 1 to tank 2. In the absence of a leak in the pipeline, the outflowq_{12} is:

$$q_{12} = \rho g(h_1(t) - h_2(t)) \tag{1}$$

In the presence of a leak, we get:

$$q_{12\ell} = \rho g(h_1(t) - h_2(t)) - q_\ell \tag{2}$$

Invoking the principle of conservation of mass, the rate of change in the volume of the tank is the difference between the inflow and outflow. Rate of change in the volume V_1 of the tank 1 is the difference between the inflow and outflow:

$$\frac{dV_1}{dt} = A_1\frac{dh_1}{dt} = q_1 - q_{12} - q_\ell \tag{3}$$

Hence, we get:

$$A_1\frac{dh_1}{dt} = q_1 - q_{12} - q_\ell \tag{4}$$

Where A_1 is the cross-sectional area of tank 1.

$$R_{12}(h_1 - h_2) = q_{12} \tag{5}$$

Where R_{12} is the resistance to flow between tank 1 and tank 2;
Using (4) and (5) yields:

$$A_1 \frac{dh_1}{dt} = q_1 - R_{12}(h_1 - h_2) - q_\ell \tag{6}$$

Similarly, the rate of change in the volume V_2 of the tank 2 becomes:

$$\frac{dV_2}{dt} = A_2 \frac{dh_2}{dt} = q_{12} - q_\ell - q_2 \tag{7}$$

$$A_2 \frac{dh_2}{dt} = q_{12} - q_\ell - q_2 \tag{8}$$

WhereA_2 is the cross-sectional area of tank 2, and q_2 is the outflow; As the flow is proportional to the pressure difference, we get:

$$R_2 h_2 = q_2 \tag{9}$$

Where R_2 is the resistance to outflow.

Remarks: In the laminar flow all the resistances, namely $R_{12} = \frac{d(h_1 - h_2)}{dq_{12}}$ in (5) and $R_2 = \frac{dh_2}{dq_2}$ in (9) are constant as the flow is laminar. For turbulent flow, these resistances are not constant and are a nonlinear function of the height:

$R_{12} = \frac{d(h_1 - h_2)}{dq_{12}} = \xi\sqrt{h_1 - h_2}$; $R_2 = \frac{dh_2}{dq_2} = \xi\sqrt{h_2}$;$\xi$ is a constant

Eliminating q_{12} and q_2 from the above Eqs. (7) and (8) yields:

$$A_2 \frac{dh_2}{dt} = R_{12}(h_1 - h_2) - R_2 h_2 - q_\ell \tag{10}$$

The continuous-time state space model is derived from (6) and (10), yields:

$$\begin{bmatrix} \frac{dh_1}{dt} \\ \frac{dh_2}{dt} \end{bmatrix} = \begin{bmatrix} -\frac{R_{12}}{A_1} & \frac{R_{12}}{A_1} \\ \frac{R_{12}}{A_2} & -\frac{R_2 + R_{12}}{A_2} \end{bmatrix} \begin{bmatrix} h_1 \\ h_2 \end{bmatrix} + \begin{bmatrix} \frac{1}{A_1} \\ 0 \end{bmatrix} q_1 - \begin{bmatrix} \frac{1}{A_1} \\ \frac{1}{A_2} \end{bmatrix} q_\ell \tag{11}$$

All the equations thus far including (11) are continuous. A linear discrete-time model is obtained by sampling the inputs and the outputs and the signals at uniformly spaced times as it is mathematically tractable, provided the time step is small-based, and there is a wealth of readily available and powerful analysis and design tools to use for such linearized models [21]. For example, the state-feedback controller based on the internal model principle, key properties of Kalman-filter-based system identification using residual model of KF, which for a linear system gives necessary and sufficient, whereas a nonlinear controller such as adaptive one provides only sufficient condition. Conventional approach based on observer, nonlinear filters, other nonlinear device cannot handle stochastic disturbance and measurement noise or gives only sufficient condition.

Closed-loop configuration: The MIMO system operates in a closed-loop configuration so the desired outputs to be regulated, denoted by $y_r(k)$, such as the height, track

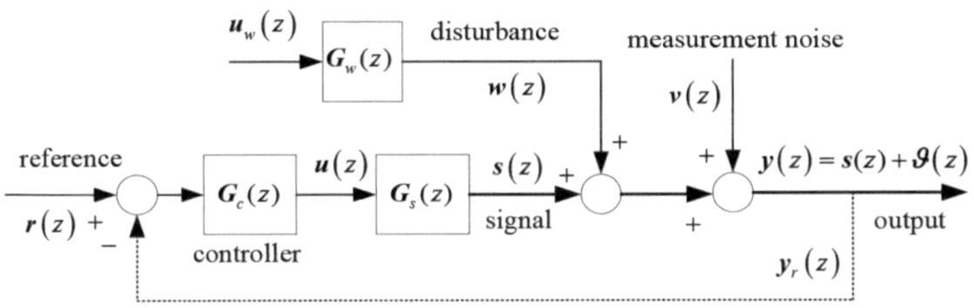

Figure 2.
Box-Jenkins model of the system relating inputs and the outputs.

the reference $\boldsymbol{y}_r(k)$ as shown in **Figure 2**. The controller $\boldsymbol{G}_c(z)$ is driven by the error between the reference and the output to be regulated $\boldsymbol{r}(z) - \boldsymbol{y}_r(z)$

$$\boldsymbol{u}(z) = \boldsymbol{G}_c(z)\left(\boldsymbol{r}(z) - \boldsymbol{y}_r(z)\right) \tag{12}$$

The signal model $\boldsymbol{G}_s(z)$ is a cascade, parallel, and feedback combination of subsystems such as the actuators, sensors, and plant.

2.1 Box-Jenkins model

Background: The Box-Jenkins method was proposed by George Box and Gwilym Jenkins in their seminal 1970 textbook Time Series Analysis: Forecasting and Control. The approach starts with the assumption that the process that generated the time series can be approximated using an ARMA model if it is stationary or an ARIMA model if it is nonstationary and comprises the following:

- *Model identification and model selection*
- *Parameter estimation* that best fit the selected ARIMA model. The most common methods use maximum likelihood estimation or nonlinear least-squares estimation.
- *Statistical model checking* by testing whether the estimated model conforms to the specifications of a stationary process

2.2 Box-Jenkins model of the proposed system

The augmented state-space representation of the system model, termed Box-Jenkins model $(\boldsymbol{A}, \boldsymbol{B}, \boldsymbol{C})$ formed of the signal model $(\boldsymbol{A}_s, \boldsymbol{B}_s, \boldsymbol{C}_s)$ and disturbance model$(\boldsymbol{A}_w, \boldsymbol{B}_w, \boldsymbol{C}_w)$ representing a p-input, q-output system [13] is given by:

$$\begin{aligned} \boldsymbol{x}(k+1) &= \boldsymbol{A}\boldsymbol{x}(k) + \boldsymbol{B}\boldsymbol{r}(k) + \boldsymbol{E}_w\boldsymbol{u}_w(k) \\ \boldsymbol{s}(k) &= \boldsymbol{C}_s\boldsymbol{x}(k) \\ \boldsymbol{y}(k) &= \boldsymbol{C}\boldsymbol{x}(k) + \boldsymbol{v}(k) \end{aligned} \tag{13}$$

$\boldsymbol{A} = \begin{bmatrix} \boldsymbol{A}_s & \boldsymbol{0} \\ \boldsymbol{0} & \boldsymbol{A}_w \end{bmatrix}; \boldsymbol{B} = \begin{bmatrix} \boldsymbol{B}_s \\ \boldsymbol{0} \end{bmatrix}; \boldsymbol{E}_w = \begin{bmatrix} \boldsymbol{0} \\ \boldsymbol{B}_w \end{bmatrix}; \boldsymbol{C} = [\boldsymbol{C}_s \;\; \boldsymbol{C}_w]; \boldsymbol{A} \in \mathbb{R}^{nxn}$ is an augmented state-transition matrix formed of $\boldsymbol{A}_s \in \mathbb{R}^{n_s x n_s}$ and $\boldsymbol{A}_w \in \mathbb{R}^{n_w x n_w}$; $\boldsymbol{B} \in \mathbb{R}^{nxp}$; $\boldsymbol{C} \in \mathbb{R}^{qxn}$;

$\boldsymbol{E}_w \in \mathbb{R}^{nxp}$ is a disturbance entry matrix; $(\boldsymbol{A}_s, \boldsymbol{B}_s, \boldsymbol{C}_s)$ and $(\boldsymbol{A}_w, \boldsymbol{B}_w, \boldsymbol{C}_w)$ are both controllable and observable. However, the signal $\boldsymbol{s}(z)$ and the disturbance $\boldsymbol{w}(z)$ may have spectral overlap. The Box-Jenkins model describes the system formed of the signal model, the disturbance model, their inputs, and the corrupted output.

Figure 2 shows the Box-Jenkins model representing the closed-loop system relating the reference $\boldsymbol{r}(z)$, the system output $\boldsymbol{y}(z)$, the output to be regulated $\boldsymbol{y}_r(z)$, the controller $\boldsymbol{G}_c(z)$, the signal model $\boldsymbol{G}_s(z)$, input $\boldsymbol{u}(z)$ to the signal model and disturbance model $\boldsymbol{G}_w(z)$, disturbance $\boldsymbol{w}(z)$ and the measurement noise $\boldsymbol{v}(z)$.

3. Kalman filter: key properties

The KF, its residual model, and key properties are restated here for the sake of completeness and convenience of the readers [12–18]. The KF associated with the fault-free unperturbed Box-Jenkins model $(\boldsymbol{A}_0, \boldsymbol{B}_0, \boldsymbol{C}_0)$ given in (13) be:

$$\begin{aligned} \hat{\boldsymbol{x}}(k+1) &= (\boldsymbol{A}_0 - \boldsymbol{K}_0\boldsymbol{C}_0)\hat{\boldsymbol{x}}(k) + \boldsymbol{B}_0\, \boldsymbol{r}(k) + \boldsymbol{K}_0\boldsymbol{y}(k) \\ \hat{\boldsymbol{y}}(k) &= \boldsymbol{C}_0\hat{\boldsymbol{x}}(k) \\ \boldsymbol{e}_{kf}(k) &= \boldsymbol{y}(k) - \hat{\boldsymbol{y}}(k) \end{aligned} \tag{14}$$

Where $\hat{\boldsymbol{x}}(k) \in R^n$ and $\hat{\boldsymbol{y}}(k)$ are respectively the best estimate of the state $\boldsymbol{x}(k)$, and of the output $\boldsymbol{y}(k)$ of the system model (13), $\boldsymbol{e}_{kf}(k)$ is the residual; the optimal Kalman gain $\boldsymbol{K}_0 \in \mathbb{R}^{6n}$ ensures the asymptotic stability of the KF, i.e. $(\boldsymbol{A}_0 - \boldsymbol{K}_0\boldsymbol{C}_0)$; $(\boldsymbol{A}_0, \boldsymbol{B}_0, \boldsymbol{C}_0)$ is the identified system model embodied in the KF. The residual model of the KF relates the residual $\boldsymbol{e}_{kf}(z)$ to the system, the desired target input $\boldsymbol{r}(z)$, and output $\boldsymbol{y}(z)$.

3.1 Key properties of the KF

The following *Lemmas* are developed here by invoking the key properties of the KF for Fault Detection and Isolation (FDI) [14]

Lemma 1:

(a) Model-matching property

The KF residual $\boldsymbol{e}_{kf}(k)$ is a zero-mean white noise process *if and only if* the identified model of the system $(\boldsymbol{A}, \boldsymbol{B}, \boldsymbol{C})$ and the true model$(\boldsymbol{A}_0, \boldsymbol{B}_0, \boldsymbol{C}_0)$ embodied in the KF are identical. This yields to:

$$\boldsymbol{e}_{kf}(k) = \boldsymbol{e}_0(k) \tag{15}$$

Where $\boldsymbol{e}_0(k)$is a zero-mean white noise process

(b) Model-mismatch property

If the identified model of the system $(\boldsymbol{A}, \boldsymbol{B}, \boldsymbol{C})$ and the true model embodied in the KF $(\boldsymbol{A}_0, \boldsymbol{B}_0, \boldsymbol{C}_0)$ are not identical, then the KF residual $\boldsymbol{e}_{kf}$ will not be a zero mean white noise process. The residual will then contain an additive term $\boldsymbol{e}_{kf}(k)$ termed fault indicator term, i.e.:

$$\boldsymbol{e}_{kf}(k) = \boldsymbol{e}_0(k) + \boldsymbol{e}_{fi}(k) \tag{16}$$

The structure of the residual model of the KF, and not that of the linear regression model of the system, is such that its equation error becomes is a colored noise process. The residual model is a function not only of the parameters of the system model, but also of the Kalman gain. The identification objective of ensuring that the KF residual is a zero-mean white noise process will ensure not only that the system model is accurately identified but also that the Kalman gain is optimal, thereby avoiding the need to specify the covariances of the disturbance and the measurement noise and to use the Riccati equation to solve for the optimal Kalman gain. The system model $(\boldsymbol{A}, \boldsymbol{B}, \boldsymbol{C})$ and its associated KF $(\boldsymbol{A} - \boldsymbol{KC}, \boldsymbol{B}, \boldsymbol{C})$, are both identified without the need for the a priori knowledge of the covariances of the disturbance and the measurement noise, by minimizing the residual of the KF [12–18]:

The identified transfer functions $\frac{\overline{\boldsymbol{D}}(z)}{\overline{F}(z)}$, and $\frac{\overline{\boldsymbol{N}}_i(z)}{\overline{F}_i(z)}$ are used to obtain the signal model, estimates of the signal $\boldsymbol{s}$, disturbance $\boldsymbol{w}(z)$ and their associated models $(\boldsymbol{A}_s, \boldsymbol{B}_s, \boldsymbol{C}_s)$ are $(\boldsymbol{A}_w, \boldsymbol{B}_w, \boldsymbol{C}_w)$ are then derived.

Lemma 2: *Signal model*

The state space model of the signal model $\boldsymbol{G}_s(z)$ or its state space representation $(\boldsymbol{A}_s, \boldsymbol{B}_s, \boldsymbol{C}_s)$ relating and the signal $\boldsymbol{s}(z)$, and desired target input $\boldsymbol{r}(k)$ are:

$$\begin{aligned} \boldsymbol{x}_s(k+1) &= \boldsymbol{A}_s\boldsymbol{x}(k) + \boldsymbol{B}_s\,\boldsymbol{r}(k) \\ \boldsymbol{s}(k) &= \boldsymbol{C}_s\boldsymbol{x}(k) \\ \boldsymbol{y}(k) &= \boldsymbol{C}\boldsymbol{x}(k) - \boldsymbol{v}(k) \\ \hat{\boldsymbol{s}}(z) &= \hat{\boldsymbol{G}}_s(z)\boldsymbol{u}(z) \end{aligned} \tag{17}$$

Where $\boldsymbol{r}(k)$ is the desired target.

Lemma 3: *Disturbance model*

The disturbance model $\boldsymbol{G}_w(z)$ or its state-space representation $(\boldsymbol{A}_w, \boldsymbol{B}_w, \boldsymbol{C}_w)$ is derived; the KF whitens the output error $\boldsymbol{\vartheta}(z) = \boldsymbol{y}(z) - \boldsymbol{s}(z)$:

$$\begin{aligned} \boldsymbol{x}_w(k+1) &= \boldsymbol{A}_w\boldsymbol{x}(k) + \boldsymbol{B}_w\,u_w(k) \\ \boldsymbol{d}(k) &= \boldsymbol{C}_s\boldsymbol{x}_w(k) \end{aligned} \tag{18}$$

3.2 Proposed KF-based scheme

- *Lemma* 1: The fault indicator term indicates the presence or absence of fault in the system model, signal model, disturbance model, or both.
- *Lemma 2:* The presence or absence of fault in the signal model is indicated.
- *Lemma 3*: The presence or absence of fault in the signal model is indicated.

3.3 Identification: emulator generated data

In view of the key properties of the KF, it is the residual model of the KF, and not the system model, that is identified in our work, thus lending our work a novelty that sets it apart from other conventional approaches. The identification objective of ensuring that the residual is a zero-mean white noise process will ensure not only that the system model is accurately identified but also that the Kalman gain is optimal, thereby avoiding the need to specify the covariance of the disturbance and the measurement noise and to use the Riccati equation to solve for the Kalman gain.

The system model, the signal model as well as the disturbance models are subject to perturbations. To account for these perturbations, emulators are connected in cascade to the output, the input, or both during the identification phase to mimic likely perturbations. It is shown in [13, 14] that the proposed identification scheme based on emulator parameter-perturbed experiments to generate likely model perturbations is superior to the conventional approach based on using either a conservative or an optimistic or no bound at all, instead of the true bound of the perturbed plant models.

In model-based, model-free, and classifier fusion approach, it is crucial to provide representative and sufficient data covering the normal, perturbed, and fault-bearing operating scenarios resulting from the variations of the subsystems. As the parameters subsystems are generally inaccessible, the data are generated indirectly by performing several off-line emulator-perturbed experiments to mimic likely operating scenarios.

In the model-based approach, the emulator-generated data are used in identification of the system and its associated Kalman filter to ensure that the identified models are robust to model-perturbation and are significantly more accurate compared with those obtained based on the classical approach of using merely the input and the output without including the perturbed models [13, 14]. System and the signal model and their associated KFs for the system and the signal models, estimation of the signal, disturbance are identified from the emulator-perturbed data.

The key properties of the KF are used to obtain the signal, the disturbance, and their models [14–18]. The fault-free Box-Jenkins system model (13) and the associated KF are identified the using the emulator-generated data by minimizing the residual $e_{kf}(z)$ to ensure the identified models are accurate, consistent, and reliable. Further, the emulator-generated data are used to provide data needed for the identification of the system, for the MFA, and the classifier fusion.

The identified transfer functions $\frac{\overline{D}(z)}{\overline{F}(z)}$, and $\frac{\overline{N}_i(z)}{\overline{F}_i(z)}$ are used to obtain the signal model, estimates of the signal s, disturbance $w(z)$, and their associated models (A_s, B_s, C_s) are (A_w, B_w, C_w) are derived.

3.4 State-feedback and feedforward controller

A block diagram of the feedforward-feedback controller implemented using the Kalman filter and internal model (A_{im}, B_{in}, C_{im}) is shown in **Figure 3** [20]. The system and signal models; Kalman filters associated with the system and signal models; feedback-feedforward controller; the signal s, the disturbance d, and the measurement noise v; residual e, estimates of the signal $\hat{s}$ and the output error $\hat{\vartheta}$ of the Kalman filter. The controller is driven by the tracking error $e_{tr}(k) = r(k) - \hat{s}(k)$.

The signal $s(z)$instead of the noisy output $y(z)$is employed for implementing the state feedback controller of the signal, desired trajectory, and output error

Feedforward controller: Even in the presence of model perturbations, the feedforward controller can mitigate the effect of the output error on the performance of the combined controller. The feedforward controller quickly rejects the output error without waiting for the deviation in the output to occur, hence its anticipatory

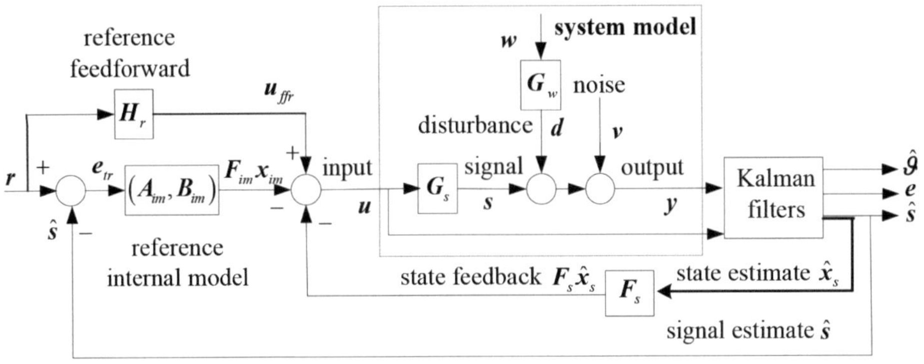

Figure 3.
Block diagram of the feedforward-feedback controller.

action. The feedforward controller of the reference, denoted by the steady-state gain $\boldsymbol{H}_r$, is the inverse signal model evaluated at the poles of the reference model. Thus:

$$\begin{aligned} \boldsymbol{H}_r &= \boldsymbol{G}_s^{-1}(z) \\ \boldsymbol{u}_{ffr} &= \boldsymbol{H}_r \boldsymbol{r}(z) \end{aligned} \tag{19}$$

3.5 Proposed Kalman filter-based scheme

The KF estimates the signal component from the output formed of an additive sum of the signal, stochastic disturbance, and measurement noise as shown.

The status of the system is asserted from the whiteness of the Kalman filter residual. A fault is detected if the residual is not a zero-mean white noise process. The faulty subsystem is isolated by estimating the perturbed emulator parameter.

When the residual is not a zero-mean white noise process signifies that either the model of the system has become faulty, i.e., a fault had occurred in the system, or the disturbance model has been perturbed, due to the purely random nature of the various disturbances affecting the system during its operation or possibly both.

3.6 Autonomous Kalman filter

Let the human operator input, u_{op} and the output y_{op} during pipeline trajectory estimation. The autonomous Kalman filter that replaces the human operator is given by:

$$\begin{aligned} \hat{\boldsymbol{x}}_{op}(k+1) &= (\boldsymbol{A}_{op} - \boldsymbol{K}_{op}\boldsymbol{C}_{op})\hat{\boldsymbol{x}}(k) + \boldsymbol{B}_{op}\, u_{op}(k) + \boldsymbol{K}_{op} y_{op}(k) \\ \hat{y}_{op}(k) &= \boldsymbol{C}_{op}\hat{\boldsymbol{x}}(k) \\ e_{op}(k) &= y_{op}(k) - \hat{y}_{op}(k) \end{aligned} \tag{20}$$

The autonomous KF drives the drone such the residual $e_{op}(k)$ is zero during entire trajectory. This will ensure that the trajectories of the human operator and KF are identical.

4. Sequential fault diagnosis

The model-free approach, namely Limit Checking and Plausibility Analysis (LCPA) and limit check, can quickly detect large faults using the limit checking, estimation of the step input responses of overshoot, rise time, and settling time. Artificial Neural Network (ANN) can capture the model of the system over both linear and nonlinear operating regimes. However, its ability to detect incipient fault critically depends upon the training data, which covers the given operating point upon the operating point, the input, and the disturbances affecting the system, the noise, and the nonlinearity effects. The same Fault Detection and Isolation (FDI) scheme may outperform other FDI schemes in some operating scenario while being outperformed by them in other scenarios. Hence, the integration of different FDI schemes will overcome this problem, in that what is missed. Relationship between these different FDI schemes will enable their collective performance to surpass that of any one of them used alone [14].

5. Model-free approaches (MFAs)

The Model-Free Approach (MFAs) are employed in monitoring the health of the system, including performance monitoring and fault detection, as it provides a macroscopic picture of the status of the system [14, 15].

5.1 Fault detection powered by the Bayesian classification for the MFA schemes

A fault in the system is asserted using Limit Checking and Plausibility (LVP) analysis from computing the step response measures of settling time, time delay, and overshoot, and using some measures based on spectral analysis, such as the frequency response and the coherence spectrum [15].

The coherence between the fault-free and actual outputs is:

$$c\left(y^0(\omega), y(\omega)\right) = \frac{\left|y^0(\omega)y(\omega)\right|^2}{\left|y^0(\omega)\right|^2\left|y(\omega)\right|^2} \tag{21}$$

Where ω is the frequency in rad/sec, and $c(y^0(\omega)y(\omega))$ are the coherence spectrum and the output of the ANN will be the fault type, i.e., either a fault in a subsystem or in a sensor. If there is no fault, then, in the ideal noise-free case, $c(y^0(\omega)y(\omega)) = 1$. Let $\hat{G}^0(\omega)$ and $\hat{G}(\omega)$ be the estimates of the frequency response of the system under normal fault-free and faulty operating regimes, respectively.

The Bayes decision strategy: The test statistic is chosen to be the median value of the coherence spectrum [14, 19]:

$$t_s = median\left\{c\left(\hat{G}^0(\omega), \hat{G}(\omega)\right)\right\} \tag{22}$$

The Bayes decision strategy used here is given by:

$$t_s \begin{cases} \leq th & for\, all\, \omega \in \Omega \quad no fault \\ > th & for\, all\, \omega \in \Omega \; fault \end{cases} \tag{23}$$

Where *th* is a threshold value, Ω is the relevant spectral region, e.g., the system bandwidth?

6. Evaluation on a physical system

The proposed sequential fault diagnosis, based on the model-based and model-free schemes, was successfully evaluated on a laboratory-scale physical process control system [7–9, 14, 15]. The controller is implemented on a two-tank process control system [22], is shown below in **Figure 4**.

6.1 Physical two-tank fluid system

The four subsystems of the two-tank system, namely the flow rate sensor γ_{s1}, the height sensor γ_{s2}, the actuator $G_1 = G_1^0\gamma_a$ where G_1^0 is fault-free, and γ_ℓ the leakage fault indicator from the from tank 1, can be affected by either a single fault or multiple ones. As shown in **Figure 3**, when either of these fault types occurs, they are detected and isolated with the integrated approach, which intelligently processes the acquired data from the various sensors, by using the most appropriate scheme (MFA or MBA) and the Bayesian classification stage to carry out the accurate and reliable fault detection and isolation. The fault-free values are $\gamma_{si} = 1 : i = 0, 1, 2, \gamma_a = 1$ and $\gamma_\ell = 1$. The net amount of outflow is $1 - \gamma_\ell$. On the testbed used, the Lab View is used for detection and isolation of faults.

Figure 5 shows the step responses of the subsystems subject to no faults, leakage faults, actuator faults, and height sensor faults. The fault magnitudes are 0.25, 0.5, and 0.75 of the fault-free cases [14, 15]. The height, flow rate, and control input profiles under various types of faults are all shown in **Figure 5**.

Subfigures A, B, and C show respectively when there is a leakage fault, an actuator fault, and a height sensor fault. Subfigures D, E, and F show respectively the effect on the flow rate of the leakage fault, actuator fault, and sensor fault, and subfigures G, H, and I show, in the same order, the effect of these same 3 faults on the control input.

The MFA approach used here includes four essential blocks, namely a limit checks, visual and plausibility block (LVP), an adaptive neuro-fuzzy inference system block (ANFIS), a fuzzy logic block (FL), and an artificial neural network block (ANN). **Figure 5** shows the effect of disturbance, measurement noise, nonlinearity, including dead-band effect, and saturation on the actuator and on the flow rate. The unwanted effect of dead-band causes delay in the system and saturation in the actuator and flow

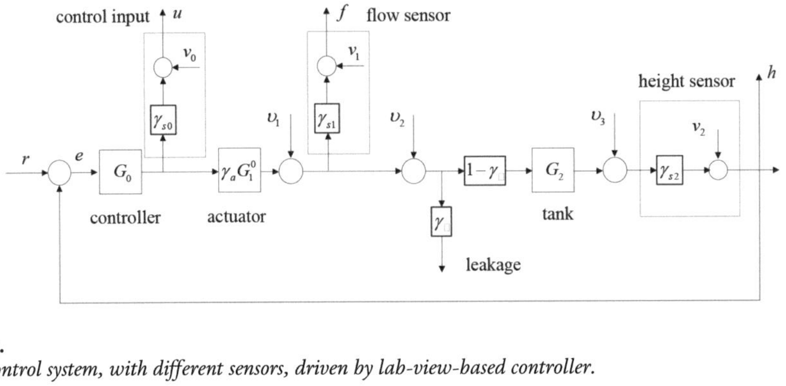

Figure 4.
Process control system, with different sensors, driven by lab-view-based controller.

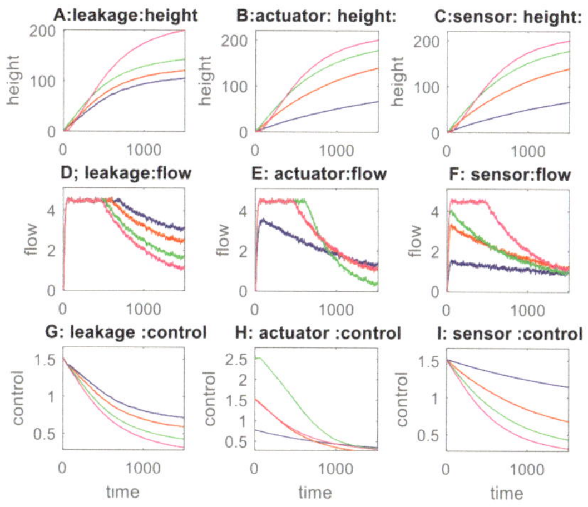

Figure 5.
Effect of leakage, actuator, and sensor faults on height, flow rate, and control: Nonlinear case.

rate. The flow rate saturates at 4.5 ml/sec. Because of the presence of the feedback in the process system, the PI controller may reduce the effect of the nonlinearity and noise if the controller is tuned accurately. However, the PI may also dissimulate the fault by rejecting it, through its loop action, as though it were a mere disturbance. This will therefore call for a careful use of the integral action in the PI controller.

6.2 Limit checking, visual, and plausibility analysis

The LPV, though limited in the size of the faults it can detect, is nevertheless the fastest of the four MFA blocks, It uses heuristics, operator experience, and the domain knowledge. It can only detect gross faults (or macro faults) but is computationally fast and monitors limit checks, flow rate, and input to the actuator for accurately determining fault status, provided the sensors are properly functioning as explained next. Some faults, such as overflow of the tanks, may not be detected or may be incorrectly reported. By way of an example, assume that the flow sensor is working properly, and the height of the tank is 250 cm, and the flow rate in the range 0 to 4 ml/sec, both of which are actual accurate values, then if the height sensor is faulty, it may then indicate an incorrect height of more than 250 cm, indicating that there is an overflow when there is none. Similarly, a faulty height sensor could report an incorrect value of 250 cm, thus indicating that the tank has reached its maximum capacity and that the sensor flow needs to be regulated to avoid an overflow, when the tank has not yet reached its full capacity. This demonstrates the weakness of the qualitative measurement of the LPV block, which, the proposed intelligent integrated approach compensates for by resorting to more accurate means of assessing the true status of the system, either through more powerful MFA blocks or through the powerful KF-based MBA block.

6.3 Artificial neural network (ANN)

ANN is a universal approximator. The height, flow rate, and control input data under leakage, actuator, and sensor faults are presented to the ANN. **Figure 4** shows the training data formed of the height, flow rate, and control input under leakage, actuator, and sensor faults, as well as the classification of the fault types.

Remarks: Except for a very few misclassifications probably due to an insufficient amount of data processed by the ANN, the estimated classes were accurate. The FDI performance of the ANN depends crucially upon the set of input-output data employed during the training phase. The training data should be sufficient in quantity and representative enough to cover all fault-free and faulty operating regimes. In practice, it is very difficult to cover all fault scenarios, especially the extreme cases involving disasters, for which data are either scarce or unavailable. The ANN approach suffers from the lack of transparency as the decision-making process is deeply embedded in the inner workings of the ANN, thus making the rationale behind the decisions taken rather unclear to the user. Nevertheless, the ANN is computationally fast and provides timely FDI [14, 15].

6.4 Fuzzy logic approach (ANFIS)

In the case of ANN and the model-based FDI scheme, the dynamic response (covering both transient and steady-state regions) of the system is presented. However, in the fuzzy logic-based approach, only the steady-state response under various operating regimes captures the benefits of both in a single framework. As such, it is regarded as a universal approximator, where the required set of fuzzy IF–THEN rules is developed automatically from the data presented to it [14, 15].

6.5 Model-based approach

The physical two-tank fluid system is nonlinear with dead-band nonlinearity. The system was identified using the emulator-based accurate and reliable scheme proposed in [13–17] wherein several offline experiments on the physical system are performed by varying the emulator parameters to reliably capture their influence on the input-output behavior. The process control system is a closed-loop single-input and multiple-output (SIMO) system relating the input $r(k)$ to the outputs, namely the control input $u(k)$, the flow rate $f(k)$, and the height $h(k)$. The system and the associated Kalman filter are identified using the prediction error method [18]. Since multiple outputs are measured, multiple Kalman filters are employed to detect and isolate the height sensor, the flow sensor, the actuator, and the leakage faults. The multiple Kalman filters included here are associated with (a) overall closed-loop system relating the input $r(k)$, to all the outputs $u(k)$, $f(k)$ and $h(k)$, (b) $u(k)$ and $f(k)$, and (c) $f(k)$ and $h(k)$.

6.6 Bayesian hypothesis testing

Fault detection is posed as a binary composite hypothesis-testing problem [7–13, 19]. The criterion to choose between the two hypotheses, namely the presence or absence of a fault, is based on minimizing the Bayes risk, which quantifies the costs associated with correct and incorrect decisions. The $Nx1$ Kalman filter residual data $\mathbf{e}(k)$ are employed for this purpose. The decision between the two hypotheses is based on comparing the likelihood ratio, which is the ratio of the conditional probabilities under the two hypotheses, to a user-defined threshold value:

$$t_s(\boldsymbol{e}) \begin{cases} \leq \eta_{th} & nofault \\ > \eta_{th} & fault \end{cases} \tag{24}$$

Where $t_s(\boldsymbol{e})$ a test statistic is computed using the residual $\boldsymbol{e}(k)$. The test statistics $t_s(\boldsymbol{e})$ depends upon the class of reference input and η_{th} is some threshold value chosen to meet the stringent and conflicting requirements of a high probability of correct fault detection and isolation with low false alarm probability.

Figure 6 shows the residuals and their test statistics, and **Figure** 7 shows the autocorrelations of the residuals when the system is subject to leakage, actuator, and sensor faults of various degrees such a small, medium, and large fault sizes. Subfigures A, B, and C; D, E, and F; and G, H, and I of **Figure 7** shows the residuals and their statistics when there is a leakage, actuator, and sensor faults, respectively. The test statistic is a constant bias of the residual, which is non-zero mean random process, and serves as an additive fault indicator term. The three sets of three subfigures each shown in **Figure** 7 namely (A, B, and C), (D, E, and F) and (G, H, and I) show the corresponding autocorrelations for different fault types.

Remarks: The test statistics indicates the fault size associated with small, medium, and large faults.

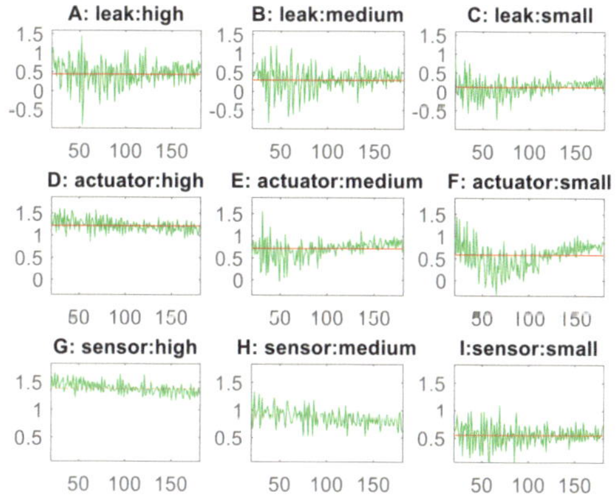

Figure 6.
The residuals and test statistics.

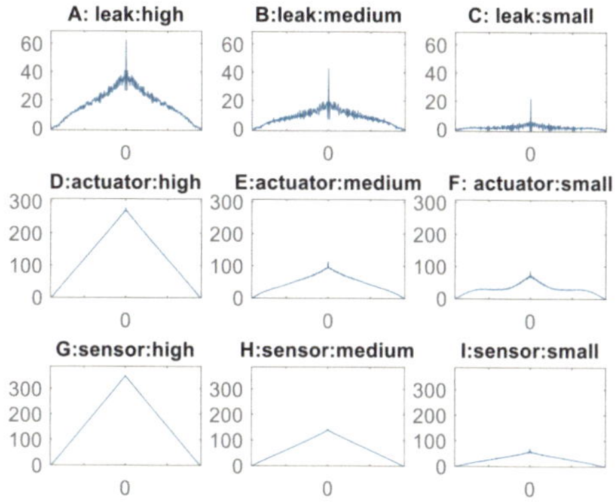

Figure 7.
Autocorrelations of the residuals.

The Bayes decision strategy was employed to assert the fault type, i.e., to decide whether it is either leakage or actuator or sensor fault, respectively, using the fault isolation scheme proposed in [13–17]. The variance of the residual, which is the maximum value of the autocorrelation function evaluated at the origin (zero delay), indicates the fault size.

7. Illustrative example

Equivalent mathematical simulation scheme with the KF and residual (purely noise) analysis should be presented, as well as numerical data (for KF tuning, including Q and R matrices). The covariances Q and R were Q=0.1 and R=1.

The estimation of the signal, the output corrupted by disturbance and measurement noise, the spectra of the signal and the disturbance is illustrated in the following simulated example [20]. The state-space model is:

$$\boldsymbol{A}_s = \begin{bmatrix} 0 & -0.7 & 0 & 0 \\ 1 & 1.5 & 0 & 0 \\ 0 & 0 & 0 & -0.8 \\ 0 & 0 & 1.7 & 0 \end{bmatrix}; \quad \boldsymbol{A}_w = \begin{bmatrix} 0.3960 & -0.8025 & 0 & 0 \\ 1 & 0 & 0 & 0 \\ 0 & 0 & 1.1326 & -0.49 \\ 0 & 0 & 1 & 0 \end{bmatrix};$$
$$\boldsymbol{B}_s = \begin{bmatrix} 0.5 & 1 \\ 1 & 0 \\ 1 & -0.3 \\ 0 & 1 \end{bmatrix}; \quad \boldsymbol{B}_w = \begin{bmatrix} 1 & 0 \\ 0 & 0 \\ 0 & 1 \\ 0 & 0 \end{bmatrix};$$
$$\boldsymbol{C} = \begin{bmatrix} 0 & 1 & 0 & 0 & 1.4160 & 0 & 0 & 0 \\ 0 & 0 & 0 & 1 & 0 & 0 & 2.9290 & 0 \end{bmatrix} \tag{25}$$

Where the order $n = 8$; $\boldsymbol{y}(k)$ is 2x1 output, $w(k)$ and $\boldsymbol{v}(k)$ are 2x1 disturbance input, and measurement noise of unity covariance zero-mean white noise processes.

Subfigures A and B at the top of **Figure 8** compare the output and the signal. Subfigures C and D show the overlapping spectra of the signal and the disturbance.

Table 1 compares the true and estimated poles of the signal and disturbances models. The estimated poles are obtained from the model reduction techniques employed in the second stage of the two-stage identification scheme

From **Table 1**, it can be deduced that identified signal and disturbance model are accurate.

Remarks: These subfigures confirm the accuracy of the estimates of the signal and the output error established in *Lemmas* 1, 2, and 3 in Section 3.

8. Kalman filter: key properties

Subfigures A and B, of **Figure 9**, compare the true step response of the signal and its Kalman filter estimate; subfigures C and D show the output error and its estimate.

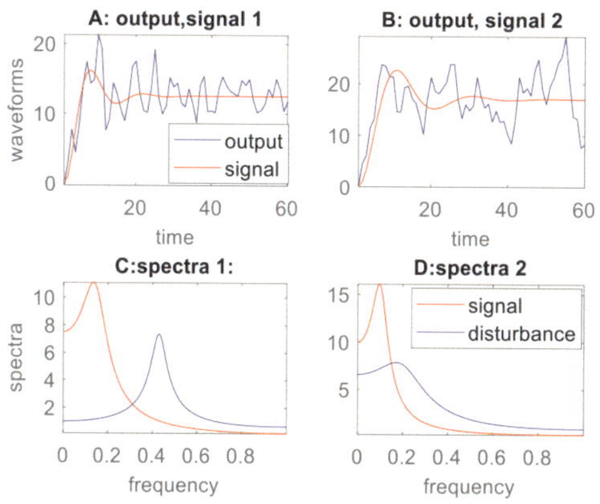

Figure 8.
Signal and its estimate; output error and its estimate.

	True poles	Identified poles
signal $\hat{G}_s(z)$	$0.7500 \pm j0.3708$ $0.8500 \pm j0.2784$	$0.7510 \pm j0.3715$ $0.8483 \pm j0.2769$
disturbance $\hat{G}_w(z)$	$0.1980 \pm j0.8737$ $0.5663 \pm j0.4114$	$0.2031 \pm j0.8752$ *Identified*

Table 1.
Poles of the signal and disturbance models.

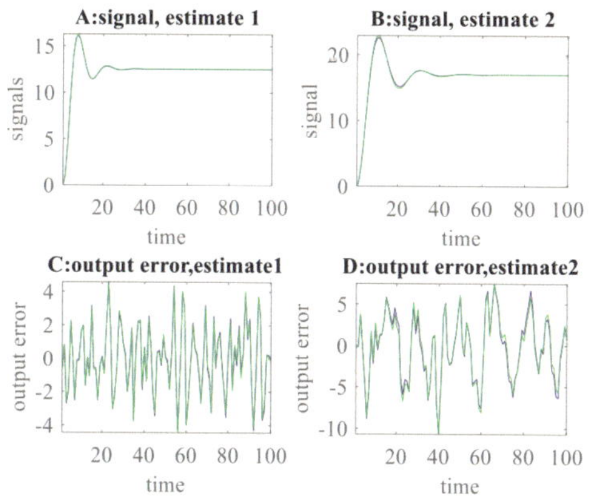

Figure 9.
Signal and output error and their estimates.

Moreover, these subfigures clearly confirm that the equation error is a colored noise, which is whitened by the KF, thus confirming key properties of *Lemmas* 1, 2, and 3. Stated in the Section 3.

Subfigures A and B, of **Figure 10** shows the autocorrelation of the equation error, whereas subfigures C and D show the autocorrelations of the residual of the Kalman filter.

Remarks: These subfigures confirm the accuracy of the estimates of the signal and the output error established in *Lemmas* 1, 2, and 3. Confirming that the equation error

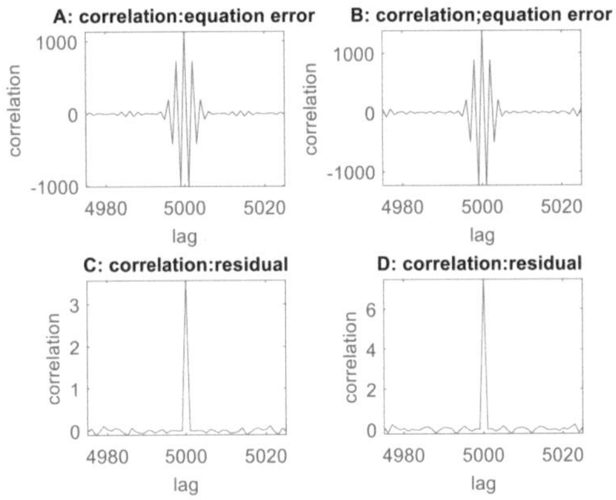

Figure 10.
Autocorrelations of KF residual and the equation error.

is a colored noise that is whitened by the KF, making the KF residual a zero-mean white noise process.

The Kalman filter-based identification of the signal and disturbance models are accurately identified the estimated poles are close to the true ones, especially those of the signal. The identification objective of ensuring the Kalman filter residual is a zero-mean white noise process will ensure not only that the system model is accurately identified but also that the Kalman gain is optimal, thereby avoiding the need to specify the covariance of the disturbance and the measurement noise and to use the Riccati equation to solve for the Kalman gain.

9. Conclusion

The novel sequential fault diagnosis approach, proposed here, is based on a judicious fusion of model-free and model-based schemes. This scheme is shown here to be superior to using (a) only the model-based scheme, (b) only the model-free scheme, or (c) the conventional combination of both schemes, in ensuring the critical requirement of a timely diagnosis and prognosis of faults with a high probability of correct decisions with a low false alarm probability. Based on extensive simulations and an evaluation on a physical system, the proposed classifier fusion scheme was shown to be reliable and efficient compared with the above-stated three conventional alternative schemes. It must be emphasized here that the novel concept of emulators and the weighted classifier scheme used here are at the core of the success of our new sequential fault diagnosis approach.

Through an integration of LPV, ANN, and FL, the model-free approach was shown to detect the presence of a possible fault quickly and reliably from the profiles of the sensor outputs. The ANN is driven by the emulator-generated data, whereas the FL is fed with steady-state values of the data. The model-free approach is also capable of providing a quick visual detection of the onset of any fault from the changes in the fault signatures such as settling time, steady-state sensor output values, and the coherence spectrum of the residuals. The fault indications obtained by the model-free approach are subsequently confirmed by the model-based approach, which, aided with a Kalman filter, provides a further necessary stage for capturing any faults, especially incipient ones, which may have escaped capture by the ANN-FL combination due either to insufficient training data or to an incomplete set of fuzzy rules. Based on extensive simulations and an evaluation on a physical system, the proposed classifier fusion scheme was shown to be reliable and efficient compared with using only a model-based or a model-free approach alone. Thanks to the emulator-based identification, the Kalman filter was shown to be accurate, reliable, and robust to modeling uncertainties including nonlinearities and neglected fast dynamics, while retaining its sensitivity to incipient faults. Further, it can perform both diagnosis and prognosis of a fault. The model-based scheme outperforms the model-free scheme in both detection and fault isolation when the system is operating in a linear region. The ANN, if presented with sufficiently representative data, is reliable in the highly nonlinear operating region. An extension of the proposed scheme to a class of nonlinear multivariable model-based scheme is currently undergoing further analysis.

Acknowledgements

Both authors acknowledge the help in obtaining some simulation data, provided, at the early stage of this research by graduate students at The University of New Brunswick and The Office of the Vice-President (Research).

Author details

Rajamani Doraiswami[1*] and Lahouari Cheded[2]

1 Department of Electrical and Computer Engineering, University of New Brunswick, Fredericton, New Brunswick, Canada

2 Life SMIEEE, UK

*Address all correspondence to: dorai@unb.ca

References

[1] Datta S, Sarkar S. Review on different pipeline fault detection methods. Science Direct Journal of Loss Prevention in the Process Industries. 2016;**41**:97-106

[2] Adegboye MA, Fung W-K, Karnik A. Recent advances in pipeline monitoring and oil leakage detection technologies: Principles and approaches. Sensors. 2019;**19**:2548

[3] Wan J et al. Hierarchical leak detection and localization method in natural gas pipeline monitoring sensor networks. Sensors. 2011;**12**(1):189-214

[4] Ziang S, Asakura T, Hayashi S. Gas leakage detection system using Kalman filter. In: 7th International Conference on Signal Processing Proceedings. IEEE Xplore. 2004;**3**:2533-2536. DOI: 10.1109/ICOSP.2004.1442297

[5] Ferrente M, Brunone B. Pipe system diagnosis and leak detection by unsteady-state tests. Advances in Water Resources. Elsevier. 2003;**26**(1):95-105

[6] An L, Sepheri N. Hydraulic actuator leak quantification using extended Kalman filter and sequential test method. In: Proceedings of American Control Conference. Minneapolis, USA: IEEE Xplore; 2006. DOI: 10.1109/ACC.2006.1657415

[7] Zadkarami M, Shahbazian M, Salahshoor K. Pipeline leak diagnosis based on wavelet and statistical features using dempster–shafer classifier fusion technique. Process Safety and Environmental Protection. 2017;**105**: 156-163

[8] Sun X, Chen T, Marquez HJ. Efficient model-based leak detection in boiler steam-water systems. Computers and Chemical Engineering. 2002;**26**: 1643-1647

[9] Tang H, Yang W, Wang Z. A model-based method for leakage detection of piston pump under variable load condition. IEEE Access. 2019;**2019**: 99771-99781

[10] Bensaad D, Soualhi A, Guillet F. A new leaky piston identification method in an axial piston pump based on the extended Kalman filter. Measurement. 2019;**148**:106921

[11] Guest Editorial. Emerging trends in LPV-based control of intelligent automotive systems. IET Control Theory & Applications. 2020;**2020**:14-18

[12] Doraiswami R, Cheded L. Linear parameter varying modelling and identification for condition-based monitoring of systems. Journal of the Franklin Institute. 2015;**352**(4): 1766-1790

[13] Doraiswami R, Cheded L. Novel Direct and Accurate Identification of Kalman filter for General Systems Described by a Box-Jenkins Model. London: Intech-Open; 2019. pp. 1-26

[14] Cheded L, Doraiswami R. A novel framework for fault diagnosis with application to process safety. Process Safety and Environmental Protection. 2021;**154**:168-171

[15] Haris KM, Doraiswami R, Cheded L. Fusion of Model-Based and Model-free Approaches to Leakage Diagnosis. Kuwait: IEEE; 2009

[16] Doraiswami R, Cheded L. Robust fault-tolerant control using an accurate emulator-based identification technique.

International Journal of Control. 2017; **2017**:1336-5820

[17] Doraiswami R, Cheded L. Robust Kalman filter-based least squares identification of a multivariable system. In: IET Control Theory and Applications: The Institution of Engineering and Technology. 2018;**12**(8). DOI: 10.1049/iet-cta.2017.0829. Available on: www.ietdl.org

[18] Doraiswami R, Cheded L. Fault tolerance in nonlinear systems: A model-based approach with a robust soft sensor design. IET Control Theory & Applications. 2021;**15**(4):499-511. DOI: 10.1049/cth2.12032. Available from: https://ietresearch.onlinelibrary.wiley.com/toc/17518652/2021/15/4 wileyonlinelibrary.com/iet-cth 499

[19] Fenton NE, Neil M. Risk Assessment and Decision Analysis with Bayesian Networks. USA: CRC Press; 2013

[20] Doraiswami R, Cheded L, Rajan S. Accurate target tracking: A New Kalman Filter Residue-based Approach Applied to a Nonlinear Multivariable Control System. In: Proceedings of the 8th International Conference of Control Systems, and Robotics (CDSR'21). Virtual Conference. May 23-25, 2021, Paper No. 303. DOI: 10.11159/cdsr21.303

[21] Proakis JG, Manolakis DG. Digital Signal Processing: Principles, Algorithms and Applications. Upper Saddle River, New Jersey: Prentice Hall; 1996

[22] Swarda K, Walvekar A, Thorat A, Mahindrakar SK. Mathematical modeling of two tank system. IJARIIE-ISSN. 2017;**3**(2):2395-4396

Chapter 4

Computationally Efficient Kalman Filter Approaches for Fitting Smoothing Splines

Joel Parker, Yifan Zhang, Bonnie J. Lafleur and Xiaoxiao Sun

Abstract

Smoothing spline models have shown to be effective in various fields (e.g., engineering and biomedical sciences) for understanding complex signals from noisy data. As nonparametric models, smoothing spline ANOVA (Analysis Of variance) models do not fix the structure of the regression function, leading to more flexible model estimates (e.g., linear or nonlinear estimates). The functional ANOVA decomposition of the regression function estimates offers interpretable results that describe the relationship between the outcome variable, and the main and interaction effects of different covariates/predictors. However, smoothing spline ANOVA (SS-ANOVA) models suffer from high computational costs, with a computational complexity of $O(N^3)$ for N observations. Various numerical approaches can address this problem. In this chapter, we focus on the introduction to a state space representation of SS-ANOVA models. The estimation algorithms based on the Kalman filter are implemented within the SS-ANOVA framework using the state space representation, reducing the computational costs significantly.

Keywords: Kalman filter, smoothing spline, functional ANOVA, state space representation, Markov structure

1. Introduction

Smoothing spline ANOVA (SS-ANOVA) has been widely used in various applications [1–3]. The representer theorem enables an exact solution of regression function in SS-ANOVA models by minimizing a regularized function in a finite-dimensional space, even though the problem resides in a infinite-dimensional space. While SS-ANOVA models have strong theoretical properties, the estimation algorithms used to fit these models are computational intensive, with a computational complexity of $O(N^3)$ for datasets with N observations. Numerous approaches have been developed to reduce the heavy computational costs of SS-ANOVA [4–7]. For example, Kim and Gu (2004) proposed to select a $q \ll N$ basis functions from N ones and reduced the computational complexity to $O(Nq^2)$. Sun et al. (2021) synergistically combined asymptotic results with the smoothing parameter estimates based on randomly selected samples with sizes of $\tilde{N} \ll N$ to reduce the computational complexity of selecting smoothing parameters to $O(\tilde{N}^3)$.

In this book chapter, we focus on the estimation approaches based on the Kalman filter. The Kalman filter was originally created to solve linear filtering and prediction problems used to generate simulations for the Apollo 11 project [8]. More recently, it has been implemented in a variety of engineering and biomedical fields [9, 10]. The Kalman filter is naturally used to fit state space models, methods that use recursive calculations on each observation entered one at a time and resulting in calculations on more accurate unknown variables after each iteration. The Kalman filter updates the state of the dynamic system given a new observation based on the state after the previous observation and the information gained from the new observation. This memory-less property and its simple recursive formulas make Kalman filter approaches computationally efficient, making them a useful tool for big data analytics. SS-ANOVA models can be reformulated to a state space representation, allowing computationally efficient Kalman filter-based model fitting and reducing computational costs to $O(N)$ for estimating univariate smoothing spline models [11]. An extension to the bivariate setting also significantly reduces the computational costs of SS-ANOVA models [12, 13].

Section 2 of this chapter will provide the theoretical background of SS-ANOVA models. Section 3 provides a brief background on state space models. The state space representation of SS-ANOVA models can be found in Section 4, along with a simulation study under the univariate setting in Section 5. Section 6 concludes this chapter.

2. Smoothing spline ANOVA models

We assume the data $(y_i, \mathbf{x}_i)$ and $i = 1,2,\dots N$ are independent and identically distributed where $y_i \in Y \in \mathbb{R}$ is the outcome/response variable and $\mathbf{x}_i \in \mathcal{X} \in \mathbb{R}^d$ represents the covariates/predictors. A nonparametric model can then be written by

$$y_i = f(\mathbf{x}_i) + e_i, \tag{1}$$

where f is a function of covariates and $e_i \sim \mathcal{N}(0, \sigma^2)$ represents the random errors. For this nonparametric model, the structure of f is not fixed and can be estimated by minimizing a penalized least squares score,

$$\frac{1}{N}\sum_{i=1}^{N} (y_i - f(\mathbf{x}_i))^2 + \lambda J(f), \tag{2}$$

where the first term measures the goodness of fit of f, and the smoothing parameter λ controls the trade-off between the goodness of fit and the roughness of f measured by $J(f)$ [2, 3, 14]. SS-ANOVA models can also handle responses from exponential families and/or correlated responses. For readers who are interested in these topics, more examples of the model estimation and implementation exist (e.g., [2, 14]). The nonparametric estimation allows f to vary in a high-dimensional (possibly infinite) space leading to more flexible results that can balance the bias-variance trade-off [2, 15]. The functional analysis of variance (ANOVA) is applied to the regression function f to improve the interpretability of model estimates by decomposing the function into main and interaction effects of covariates. These main and interaction effects can be estimated in the corresponding subspaces of the reproducing kernel Hilbert space (RKHS), which is introduced in the next section.

2.1 ANOVA

2.1.1 Classical ANOVA

Classical ANOVA can be used to help to understand the decomposition of regression function in (1). We use a one-way classical ANOVA model as an example. The outcome y_i can be modeled by $y_{ij} = \mu_i + e_{ij}$, where μ_i is the mean treatment levels with $i = 1,2,\cdots,K_1$ and $j = 1,2,\cdots,K_2$. The terms in this model can be rewritten as

$$y_{ij} = \mu + \delta_i + e_{ij}, \tag{3}$$

where μ is the overall mean effect and δ_i is the treatment effect. Side conditions are added to ensure the uniqueness of this decomposition. Now consider the univariate nonparametric function in (1). The regression function can be written as

$$f(x) = Af + (I - A)f = f_0 + f_1 \tag{4}$$

where A is an averaging operator that averages the effect of x and I is the identity operator. We also need to add some side conditions for this decomposition to ensure the uniqueness of the decomposition of the regression function.

2.1.2 Functional ANOVA

The multivariate function $f(x_{\langle 1\rangle}, x_{\langle 2\rangle}, \ldots, x_{\langle d\rangle})$ on a d-dimensional product domain $\mathcal{X} = \prod_{j=1}^{d} \mathcal{X}_j \in \mathcal{R}^d$ can be decomposed similarly to the classical ANOVA in the RKHS. The construction of the RKHS on $\Pi_{j=1}^{d} \mathcal{X}_j$ is by taking the tensor product over the marginal domains $\mathcal{X}_j$. We need the following theorem to construct the tensor-product space.

Theorem 1.1 If $R_1(x_{\langle 1\rangle}, \tilde{x}_{\langle 1\rangle})$ is nonnegative definite on $\mathcal{X}_1$ and $R_2(x_{\langle 2\rangle}, \tilde{x}_{\langle 2\rangle})$ is nonnegative definite on $\mathcal{X}_2$, then $R(x_{\langle 1\rangle}, x_{\langle 2\rangle}) = R_1(x_{\langle 1\rangle}, \tilde{x}_{\langle 1\rangle})R_2(x_{\langle 2\rangle}, \tilde{x}_{\langle 2\rangle})$ is nonnegative definite on $\mathcal{X} = \mathcal{X}_1 \times \mathcal{X}_2$.

Theorem 1.1 implies that the RKHS $\mathcal{H}$ on $\Pi_{j=1}^{d} \mathcal{X}_j$ has the reproducing kernel $R = \Pi_{j=1}^{d} R_{\langle j\rangle}$, where $R_{\langle j\rangle}$ is the reproducing kernel for $\mathcal{H}_{\langle \gamma\rangle}$ on $\mathcal{X}_{\langle \gamma\rangle}$. Additionally, the Hilbert space $\mathcal{H}_{\langle j\rangle}$ can be decomposed into $\mathcal{H}_{\langle j\rangle} = \mathcal{H}_{\langle j\rangle(0)} \oplus \mathcal{H}_{\langle j\rangle(1)}$, where $\mathcal{H}_{\langle j\rangle(0)}$ is the null space and $\mathcal{H}_{\langle j\rangle(1)}$ is the orthogonal complement to $\mathcal{H}_{\langle j\rangle(0)}$. Then $\mathcal{H} = \otimes_{j=1}^{d} \mathcal{H}_{\langle j\rangle}$ can be decomposed as

$$\begin{aligned} \mathcal{H} &= \otimes_{j=1}^{d} (\mathcal{H}_{\langle j\rangle(0)} \oplus \mathcal{H}_{\langle j\rangle(1)}) \\ &= \oplus_{\mathcal{S}} \{ (\otimes_{j \in \mathcal{S}} \mathcal{H}_{\langle j\rangle(1)}) \otimes (\otimes_{j \notin \mathcal{S}} \mathcal{H}_{\langle j\rangle(0)}) \} \\ &= \oplus_{\mathcal{S}} \mathcal{H}_{\mathcal{S}} \end{aligned} \tag{5}$$

where $\mathcal{S}$ denotes all of the subsets of $\{1, \ldots, d\}$. The term $\mathcal{H}_{\mathcal{S}}$ has the reproducing kernel $R_{\mathcal{S}} \propto \Pi_{j \in \mathcal{S}} R_{\langle j\rangle(1)}$.

In general, the inner product in $\mathcal{H}$ can be specified as

$$J(f, g) = \sum_{j=1}^{B} \theta_j^{-1} \langle f_j, g_j \rangle_j \tag{6}$$

where $\theta_j \geq 0$ are tuning parameters and B is the number of smoothing parameters. The roughness penalty in (2) can be written in the form of (6). Then the reproducing kernel associated with (6) can be written as

$$R = \sum_{j=1}^{B} \theta_j R_j, \tag{7}$$

where R_j is the reproducing kernel for the corresponding tensor product RKHS in (5).

2.2 An example of RKHS on $[0, 1]^2$

We will use one example of an RKHS on $[0, 1]^2$ to demonstrate the decomposition of RKHS. More examples of discrete and/or continuous domains can be found in Gu (2013) [2]. We consider the following tensor sum decomposition on $[0, 1]$,

$$\begin{aligned} \mathcal{H}_{\langle 1 \rangle} &= \left\{ f : \int_0^1 \left(f^{(2)}(x) \right)^2 dx < \infty \right\} \\ &= \{ f : f \propto 1 \} \oplus \{ f : f \propto k_1(x) \} \\ &\oplus \left\{ f : \int_0^1 f dx = \int_0^1 f^{(1)} dx = \int_0^1 f^{(2)} dx = 0, f^{(2)} \in \mathcal{L}_2[0, 1] \right\} \\ &= \mathcal{H}_{\langle 1 \rangle (00)} \oplus \mathcal{H}_{\langle 1 \rangle (01)} \oplus \mathcal{H}_{\langle 1 \rangle (1)}, \end{aligned} \tag{8}$$

where $\mathcal{H}_{\langle 1 \rangle 01} \oplus \mathcal{H}_{\langle 1 \rangle 1}$ forms the contrast in the one-way ANOVA decomposition, and the function k_r is a scaled Bernoulli polynomial function with $k_r(x) = B_r(x)/r!$ [2, 16]. The RKHS has three reproducing kernels $R_{\langle 1 \rangle 00}(x, \tilde{x}) = 1$, $R_{\langle 1 \rangle 01}(x, \tilde{x}) = k_1(x) k_1(\tilde{x})$, and $R_{\langle 1 \rangle 1}(x, \tilde{x}) = k_2(x) k_2(\tilde{x}) - k_4(|x - \tilde{x}|)$.

Now consider the RKHS $\mathcal{H}$ on $[0, 1] \times [0, 1]$. Here $\mathcal{H}$ can be the tensor product spaces of $\mathcal{H}_{\langle 1 \rangle}$ on $[0, 1]$ and $\mathcal{H}_{\langle 2 \rangle}$ on $[0, 1]$. Based on the tensor sum decomposition in (8), we have

$$\begin{aligned} \mathcal{H} &= \mathcal{H}_{\langle 1 \rangle} \otimes \mathcal{H}_{\langle 2 \rangle} \\ &= \left(\mathcal{H}_{\langle 1 \rangle (00)} \oplus \mathcal{H}_{\langle 1 \rangle (01)} \oplus \mathcal{H}_{\langle 1 \rangle (1)} \right) \otimes \left(\mathcal{H}_{\langle 2 \rangle (00)} \oplus \mathcal{H}_{\langle 2 \rangle (01)} \oplus \mathcal{H}_{\langle 2 \rangle (1)} \right) \\ &= \left(\mathcal{H}_{\langle 1 \rangle (00)} \otimes \mathcal{H}_{\langle 2 \rangle (00)} \right) \oplus \left(\mathcal{H}_{\langle 1 \rangle (00)} \otimes \mathcal{H}_{\langle 2 \rangle (01)} \right) \oplus \left(\mathcal{H}_{\langle 1 \rangle (01)} \otimes \mathcal{H}_{\langle 2 \rangle (00)} \right) \\ &\oplus \left(\mathcal{H}_{\langle 1 \rangle (01)} \otimes \mathcal{H}_{\langle 2 \rangle (01)} \right) \oplus \left(\mathcal{H}_{\langle 1 \rangle (00)} \otimes \mathcal{H}_{\langle 2 \rangle (1)} \right) \oplus \left(\mathcal{H}_{\langle 1 \rangle (1)} \otimes \mathcal{H}_{\langle 2 \rangle (00)} \right) \\ &\oplus \left(\mathcal{H}_{\langle 1 \rangle (01)} \otimes \mathcal{H}_{\langle 2 \rangle (1)} \right) \oplus \left(\mathcal{H}_{\langle 1 \rangle (1)} \otimes \mathcal{H}_{\langle 2 \rangle (01)} \right) \oplus \left(\mathcal{H}_{\langle 1 \rangle (1)} \otimes \mathcal{H}_{\langle 2 \rangle (1)} \right), \end{aligned} \tag{9}$$

where the first four terms in (9) are in the null space $\mathcal{H}_{(\mathbf{0})}$ and the remaining five terms are in the orthogonal complement (i.e., $\mathcal{H}_{(\mathbf{1})}$). The reproducing kernel for the cubic spline (m=2) for each subspace can be found in **Table 1**.

Subspace	Reproducing Kernel
$\mathscr{H}_{\langle 1\rangle(00)} \otimes \mathscr{H}_{\langle 2\rangle(00)}$	1
$\mathscr{H}_{\langle 1\rangle(01)} \otimes \mathscr{H}_{\langle 2\rangle(00)}$	$k_1(x_{\langle 1\rangle})k_1(\tilde{x}_{\langle 1\rangle})$
$\mathscr{H}_{\langle 1\rangle(01)} \otimes \mathscr{H}_{\langle 2\rangle(01)}$	$k_1(x_{\langle 1\rangle})k_1(\tilde{x}_{\langle 1\rangle})k_1(x_{\langle 2\rangle})k_1(\tilde{x}_{\langle 2\rangle})$
$\mathscr{H}_{\langle 1\rangle(1)} \otimes \mathscr{H}_{\langle 2\rangle(00)}$	$k_2(x_{\langle 1\rangle})k_2(\tilde{x}_{\langle 1\rangle}) - k_4(\lvert x_{\langle 1\rangle} - \tilde{x}_{\langle 1\rangle}\rvert)$
$\mathscr{H}_{\langle 1\rangle(1)} \otimes \mathscr{H}_{\langle 2\rangle(01)}$	$[k_2(x_{\langle 1\rangle})k_2(\tilde{x}_{\langle 1\rangle}) - k_4(\lvert x_{\langle 1\rangle} - \tilde{x}_{\langle 1\rangle}\rvert)]k_1(x_{\langle 2\rangle})k_1(\tilde{x}_{\langle 2\rangle})$
$\mathscr{H}_{\langle 1\rangle(1)} \otimes \mathscr{H}_{\langle 2\rangle(1)}$	$[k_2(x_{\langle 1\rangle})k_2(\tilde{x}_{\langle 1\rangle}) - k_4(\lvert x_{\langle 1\rangle} - \tilde{x}_{\langle 1\rangle})]\,[k_2(x_{\langle 2\rangle})k_2(\tilde{x}_{\langle 2\rangle}) - k_4(\lvert x_{\langle 2\rangle} - \tilde{x}_{\langle 2\rangle}\rvert)]$

Table 1.
Subspaces and their corresponding reproducing kernels for the RKHS, $\mathscr{H}$, on $[0, 1] \times [0, 1]$.

2.3 Estimation

The estimation algorithm of SS-ANOVA models relies on the following representer theorem.

Theorem 1.2 (Representer Theorem) There exist coefficient vectors $\boldsymbol{\xi} = (\xi_1, \ldots, \xi_M)' \in \mathcal{R}^M$ and $\mathbf{c} = (c_1, \ldots, c_N)' \in \mathcal{R}^N$ such that the minimizer of (2) in $\mathcal{H} = \mathcal{H}_0 \oplus \mathcal{H}_1$ has the following representation:

$$f(\mathbf{x}) = \sum_{m=1}^{M} \xi_m \phi_m(\mathbf{x}) + \sum_{i=1}^{N} c_i R(\mathbf{x}_i, \mathbf{x}), \tag{10}$$

where $\{\phi_m, m = 1, \ldots, M\}$ are the basis functions of the null space $\mathcal{H}_0$ and $R(\cdot, \cdot)$ is the reproducing kernel of $\mathcal{H}_1$.

Taking into consideration model (1), the function f can be estimated by minimizing (2). Using the representer theorem, the function f can be written as

$$\mathbf{f} = \mathbf{S}\boldsymbol{\xi} + \mathbf{Q}\mathbf{c} \tag{11}$$

where $\mathbf{f} = (f(\mathbf{x}_1), \ldots, f(\mathbf{x}_N))'$, $\mathbf{S}$ is a $N \times M$ matrix (e.g., $M = 2$ for cubic spline) where the $(i, j)^{th}$ entry is $\phi_j(\mathbf{x}_i)$ and $\mathbf{Q}$ is a $N \times N$ matrix where the $(i, j)^{th}$ entry is $R(\mathbf{x}_i, \mathbf{x}_j)$ of the form (7). Plugging (11) into (2), the penalized least squares can be written as

$$\frac{1}{N}(\mathbf{y} - \mathbf{S}\boldsymbol{\xi} - \mathbf{Q}\mathbf{c})'(\mathbf{y} - \mathbf{S}\boldsymbol{\xi} - \mathbf{Q}\mathbf{c}) + \lambda \mathbf{c}'\mathbf{Q}\mathbf{c} \ . \tag{12}$$

We differentiate (12) with respect to $\boldsymbol{\xi}$ and $\mathbf{c}$ to obtain the linear system

$$\begin{aligned} (\mathbf{Q} + N\lambda\mathbf{I})\mathbf{c} + \mathbf{S}\boldsymbol{\xi} &= \mathbf{y}, \\ \mathbf{S}'\mathbf{c} &= 0. \end{aligned} \tag{13}$$

For given smoothing parameters, solving for $\mathbf{c}$ and $\boldsymbol{\xi}$ provides the estimation for f. The selection of smoothing parameters for SS-ANOVA models is introduced below.

2.4 Selection of smoothing parameters

The smoothing parameters $\lambda/\boldsymbol{\theta}$ balance the trade-off between the goodness of fit for f and the roughness of f in (2). Choosing the optimal smoothing parameters is data specific and should be performed prior to nonparametric regression analysis. Several smoothing parameter selection methods have been developed for the SS-ANOVA models [17]. Generalized cross-validation (GCV) is one of the most popular methods for selecting the optimal smoothing parameters $\lambda/\boldsymbol{\theta}$ [18, 19].

To avoid overparameterization, let $\boldsymbol{\lambda} = (\lambda/\theta_1, \ldots \lambda/\theta_B)'$. The GCV score is defined as

$$V(\boldsymbol{\lambda}) = \frac{N^{-1}\mathbf{y}'(I - A(\boldsymbol{\lambda}))^2\mathbf{y}}{\left[N^{-1}tr(I - \alpha A(\boldsymbol{\lambda}))\right]^2} \tag{14}$$

where $A(\boldsymbol{\lambda})$ is symmetric matrix similar to the hat matrix in linear regression, and $tr(\cdot)$ represents trace. The parameter $\alpha \geq 1$ is a fudge factor [4]. When $\alpha = 1$ it is the original GCV score. Larger α's yield smoother estimates. By default, we set $\alpha = 1.4$. Then optimal smoothing parameters $\boldsymbol{\lambda}$ can be chosen by minimizing the GCV score (14) using Newton-Raphson methods.

2.5 Computational complexity

In this section, we will discuss the computation complexity for calculating $\mathbf{c}$ and $\boldsymbol{\xi}$ from (12). One requires $N^3/3 + O(N^2)$ operations to obtain estimates of $\mathbf{c}$ and $\boldsymbol{\xi}$ for the fixed smoothing parameters. In practice, the optimization of the smoothing parameter is also needed, which requires operations of $4BN^3/3 + O(N^2)$, where B is the number of smoothing parameters. Therefore, to minimize (12), the estimation algorithms have a computational complexity of $O(N^3)$. The following sections will discuss how the Kalman filter can be used to fit SS-ANOVA models, which reduces the computation complexity to $O(N)$.

3. State space models

State space methodology was traditionally used to study dynamic problems (e.g., space tracking settings) because the procedure allows for "real-time" updating as data are collected [20]. In this chapter, we use the linear Gaussian state space model as an example to introduce concepts of state space models using the Kalman filter approach. More applications of state space models can be found in a study by Douc, Moulines and Stoffer (2014) and Durbin and Koopman (2001) [21, 22]. A state space model consists of two equations: state equation and measurement equation. The state equation describes the dynamics of the state variables:

$$\mathbf{z}_{t+1} = \mathbf{G}_t\mathbf{z}_t + \Psi_t\boldsymbol{\eta}_t, \qquad \boldsymbol{\eta}_t \overset{\text{iid}}{\sim} \mathcal{N}(0, \Delta_t), \tag{15}$$

where $\mathbf{z}_t$ is the $h \times 1$ state vector, $\boldsymbol{\eta}_t$ is the $g \times 1$ disturbance vector with zero mean and a covariance matrix Δ_t, and $\mathbf{G}_t$ and Ψ_t are fixed design matrices of dimensions

$h \times h$ and $h \times g$, respectively. The measurement equation shows the relationship between the observed variable and the unobserved state variable:

$$\mathbf{o}_t = \Phi_t \mathbf{z}_t + \boldsymbol{\varepsilon}_t, \qquad \boldsymbol{\varepsilon}_t \overset{\text{iid}}{\sim} \mathcal{N}(0, \mathbf{V}_t), \tag{16}$$

where $\mathbf{o}_t$ is a $n \times 1$ vector with n observations, $\boldsymbol{\varepsilon}_t$ is a $n \times 1$ disturbance vector with zero mean and a covariance matrix $\mathbf{V}_t$, and Φ_t is a fixed design matrix of dimension $n \times h$. The initial state vector $\mathbf{z}_0$ is assumed to be normally distributed with mean μ_0 and covariance matrix $\mathbf{P}$. The two vectors $\boldsymbol{\eta}_t$ and $\boldsymbol{\varepsilon}_t$ are assumed to be mutually uncorrelated, that is,

$$\begin{pmatrix} \boldsymbol{\eta}_t \\ \boldsymbol{\varepsilon}_t \end{pmatrix} \overset{\text{iid}}{\sim} \mathcal{N}\left[\mathbf{0}, \begin{pmatrix} \Delta_t & \mathbf{0} \\ \mathbf{0} & \mathbf{V}_t \end{pmatrix}\right]. \tag{17}$$

These two vectors are also uncorrelated to the initial state vector $\mathbf{z}_0$.

The Kalman filter utilizes a set of recursive equations to estimate $\mathbf{z}_t$, given the observations $\mathbf{O}_t = \{\mathbf{o}_1, \ldots, \mathbf{o}_t\}$ at time t and its error variance matrix $\mathbf{P}_t$ [13]. Define

$$\begin{aligned} \hat{\mathbf{z}}_t &= \mathbb{E}[\mathbf{z}_t | \mathbf{O}_t], \\ \mathbf{P}_t &= \mathbb{E}\left[(\mathbf{z}_t - \hat{\mathbf{z}}_t)\left((\mathbf{z}_t - \hat{\mathbf{z}}_t)' | \mathbf{O}_t\right],\right. \end{aligned} \tag{18}$$

where $\hat{\mathbf{z}}_t$ is the Kalman filter estimation of $\mathbf{z}_t$, with $\mathbf{z}_0 = \mathbb{E}[\mathbf{z}_0] = \mu_0$, and $\mathbf{P}_0 = \mathbf{P}$. From the state and measurement Eqs. (15) and (16), the estimated $\mathbf{z}_{t|t-1}$ and the covariance matrix given $\mathbf{z}_{t-1}$, and $\mathbf{P}_{t-1}$ become

$$\begin{aligned} \hat{\mathbf{z}}_{t|t-1} &= \mathbb{E}[\mathbf{z}_t | \mathbf{O}_{t-1}] = \mathbf{G}_t \mathbf{z}_{t-1}, \\ \mathbf{P}_{t|t-1} &= \mathbb{E}\left[(\mathbf{z}_t - \hat{\mathbf{z}}_{t|t-1})\left((\mathbf{z}_t - \hat{\mathbf{z}}_{t|t-1})' | \mathbf{O}_{t-1}\right] = \mathbf{G}_t \mathbf{P}_{t-1} \mathbf{G}_t' + \Psi_t \Delta_t \Psi_t',\right. \end{aligned} \tag{19}$$

and

$$\mathbf{o}_{t|t-1} = \Phi_t \ \hat{\mathbf{z}}_{t|t-1}. \tag{20}$$

Then the prediction error vector $\mathbf{v}_t$ is

$$\mathbf{v}_t = \mathbf{o}_t - \mathbf{o}_{t|t-1} = \mathbf{o}_t - \Phi_t \ \hat{\mathbf{z}}_{t|t-1} = \Phi_t(\mathbf{z}_t - \hat{\mathbf{z}}_{t|t-1}) + \boldsymbol{\varepsilon}_t, \tag{21}$$

with the covariance matrix

$$\mathbb{E}\left[\mathbf{v}_t \mathbf{v}_t'\right] = \Lambda_t = \Phi_t \mathbf{P}_{t|t-1} \Phi_t' + \mathbf{V}_t. \tag{22}$$

Using the facts of the joint distribution of $\mathbf{z}_t$ and $\mathbf{v}_t$, the Kalman filter estimator $\hat{\mathbf{z}}_t = \mathbb{E}(\mathbf{z}_t | \mathbf{O}_t)$ at time t and its covariance matrix can be updated using

$$\begin{aligned} \hat{\mathbf{z}}_t &= \hat{\mathbf{z}}_{t|t-1} + \mathbf{P}_{t|t-1} \Phi_t' \Lambda_t^{-1}(\mathbf{o}_t - \Phi_t \hat{\mathbf{z}}_{t|t-1}) \\ &= \mathbf{G}_t \mathbf{z}_{t-1} + \mathbf{P}_{t|t-1} \Phi_t' \Lambda_t^{-1} \mathbf{v}_t, \\ \mathbf{P}_t &= \mathbf{P}_{t|t-1} - \mathbf{P}_{t|t-1} \Phi_t' \Lambda_t^{-1} \Phi_t \mathbf{P}_{t|t-1}. \end{aligned} \tag{23}$$

We further define Kalman gain as

$$\mathbf{K}_t = \mathbf{P}_{t|t-1}\Phi_t'\Lambda_t^{-1}. \tag{24}$$

Then the filtered estimate of $\mathbf{z}_t$ and its covariance matrix $\mathbf{P}_t$ is

$$\begin{aligned}\hat{\mathbf{z}}_t &= \hat{\mathbf{z}}_{t|t-1} + \mathbf{K}_t\mathbf{v}_t,\\ \mathbf{P}_t &= (\mathbf{I} - \mathbf{K}_t\Phi_t)\mathbf{P}_{t|t-1}.\end{aligned} \tag{25}$$

4. State space representation of SS-ANOVA models

Due to the Markov structure of SS-ANOVA models after reparameterization, the SS-ANOVA models can be represented by the state space models, allowing for efficient estimation by algorithms based on the Kalman filter. Wecker and Ansley (1983) showed the state space representation for univariate smoothing spline models [11]. Such an approach reduces the computational complexity of smoothing splines from $O(N^3)$ to $O(N)$. Based on the fast algorithm for the multivariate Kalman filter, Qin and Guo (2006) extended the univariate case to multivariate SS-ANOVA models [12]. The two-dimensional procedure was implemented, with computational complexity of $O(n_1n_2^3)$ for data of size $N = n_1n_2$. The extension to higher dimensions was also discussed. In this chapter, we focus on the univariate setting to demonstrate the procedure to derive the state space representation of smoothing splines.

4.1 Univariate setting

We can use the state space formulation to represent model (1) $(d = 1)$ and apply the Kalman filter algorithm to estimate the model parameters of smoothing spline models efficiently. Based on the pioneered work in Wahba (1978) [23], the univariate function $f(x)$ can be written in the following form

$$f(x) = \sum_{\nu=0}^{m-1} \alpha_\nu \frac{(x - x_l)^\nu}{\nu!} + \sqrt{\lambda}\sigma \int_{x_l}^{x} \frac{(x - h)^{m-1}}{(m-1)!} dW(h), \tag{26}$$

where the covariate $x \in [x_l, x_u]$ and $W(h)$ is a Wiener process with the unit dispersion parameter. When all α's have the diffuse prior distribution, the conditional expectation of $f(x)$ given all data is the function minimizing (2) with the smoothing parameter $1/\lambda$. The model (26) can be rewritten as

$$y_i = \mathbf{1}'\mathbf{U}(x_i) + e_i, \quad i = 1,\cdots,N, \tag{27}$$

where $\mathbf{1}' = [1, 0, \cdots, 0]$ and $\mathbf{U}(x) = \left[U^{(m)}(x), \cdots, U^{(1)}(x)\right]'$ is the m-dimensional stochastic process. In particular, we define

$$U^{(j)}(x) = \sum_{\nu=0}^{j-1} U^{(m-\nu)}(x_l) \frac{(x - x_l)^\nu}{\nu!} + \sqrt{\lambda}\sigma \int_{x_l}^{x} \frac{(x - h)^{m-1}}{(m-1)!} dW(h), \quad j = m,\cdots,1. \tag{28}$$

The vector $\mathbf{U}$ contains $U^{(m)}(x)$ and its first $(m-1)$ derivatives. Let $\alpha_\nu = U^{(m-\nu)}(x_l)$, and we can easily verify the model (27).

The state space formulation relies on the Markov structure of $\mathbf{U}(x)$, which is demonstrated below. We define a $m \times m$ matrix, $\Gamma_m(x_b, x_a)$, for any x_b and x_a within the interval $[x_l, x_u]$ as

$$\Gamma_m(x_b, x_a) = \begin{pmatrix} 1 & (x_b - x_a) & \dots & \dfrac{(x_b - x_a)^{m-1}}{(m-1)!} \\ & 1 & \dots & \dfrac{(x_b - x_a)^{m-2}}{(m-2)!} \\ & & & \vdots \\ & & & 1 \end{pmatrix}. \tag{29}$$

We can easily verify

$$\Gamma_m(x_c, x_a) = \Gamma_m(x_c, x_b)\Gamma_m(x_b, x_a). \tag{30}$$

To show the Markov structure of $\mathbf{U}(x)$, we also define a $m \times 1$ random vector $\omega(x_b, x_a) = \left[\omega^{(1)}(x_b, x_a), \cdots, \omega^{(m)}(x_b, x_a)\right]'$, where

$$\omega^{(\nu)}(x_b, x_a) = \sqrt{\lambda}\sigma \int_{x_a}^{x_b} \frac{(x_b - h)^{\nu-1}}{(\nu-1)!} dW(h), \quad \nu = 1,\cdots,m. \tag{31}$$

For any $\nu = 1,\cdots,m$, we have

$$\omega^{(\nu)}(x_c, x_a) = \omega^{(\nu)}(x_c, x_b) + \sum_{j=0}^{\nu-1} \frac{(x_c - x_b)^j}{j!} \omega^{(\nu-j)}(x_b, x_a). \tag{32}$$

Thus we have

$$\omega(x_b, x_a) = \Gamma_m(x_c, x_b)\boldsymbol{\omega}(x_b, x_a) + \boldsymbol{\omega}(x_c, x_b). \tag{33}$$

We now apply (30) and (33) to obtain the Markov structure of $\mathbf{U}$

$$\begin{aligned} \mathbf{U}(x_b) &= \Gamma_m(x_b, x_l)\mathbf{U}(x_l) + \omega(x_b, x_l) \\ &= \Gamma_m(x_b, x_a)\Gamma_m(x_a, x_l)\mathbf{U}(x_l) + \Gamma_m(x_b, x_a)\boldsymbol{\omega}(x_a, x_l) + \boldsymbol{\omega}(x_b, x_a) \\ &= \Gamma_m(x_b, x_a)\mathbf{U}(x_a) + \boldsymbol{\omega}(x_b, x_a). \end{aligned} \tag{34}$$

The Markov structure is the key to the state space representation of SS-ANOVA models. For $x_l = x_1 \leq \cdots \leq x_n = x_u$, we have

$$y_i = \sum_{\nu=0}^{m-1} \alpha_\nu \frac{(x_i - x_l)^\nu}{\nu!} + \mathbf{1}'\Omega(x_i) + e_i, \tag{35}$$

as the measurement equation, where $\Omega(x_i) = \left[\Omega^{(m)}(x_i), \cdots, \Omega^{(1)}(x_i)\right]'$ and $\Omega^{(\nu)}(x_i) = \omega^{(\nu)}(x_i, x_l)$ for $\nu = 1,\cdots,m$ and $i = 1,\cdots,N$. From (33), we have the state equation

$$\Omega(x_i) = \Gamma_m(x_i, x_{i-1})\Omega(x_{i-1}) + \boldsymbol{\omega}(x_i, x_{i-1}), \tag{36}$$

for $i = 1,\cdots,N$. Given the parameters λ and σ, the Kalman filtering and smoothing algorithms can be implemented to perform estimation for this state space representation.

5. Simulation studies

We used simulated data to compare the estimates of smoothing splines with those based on state space representation under the univariate setting. The following model was used to simulate $N = 1{,}000$ observations.

$$y_i = 7\sin(\pi^* x_i) + e_i, \tag{37}$$

where $x_i = t_i/100$, $t_i = 1,\cdots,1{,}000$, and $e_i \sim \mathcal{N}(0, 1)$. To apply the univariate SS-ANOVA model to the simulated data, we used the *ssanova* function in the *gss* package (version 2.2-3) [24]. The GCV algorithm was used to select the smoothing parameter of SS-ANOVA models. To implement the Kalman filtering algorithm to fit the smoothing splines, we used Eq. (35) as the measurement equation and Eq. (36) as the state equation. The parameters λ and σ were set to 0.01 and 1, respectively. We fitted the cubic spline (i.e., $m = 2$) in the simulation studies. In the state Eq. (36), we have

$$\Gamma_2(x_i, x_{i-1}) = \begin{pmatrix} 1 & (x_i - x_{i-1}) \\ 0 & 1 \end{pmatrix}, \tag{38}$$

for $i = 2,\cdots,1{,}000$. The $\nu\nu'$th element of the variance matrix of $\boldsymbol{\omega}(x_i, x_{i-1})$ is

$$\lambda\sigma^2 \frac{(x_i - x_{i-1})^{\nu+\nu'-1}}{(\nu+\nu'-1)(\nu-1)!(\nu'-1)!}, \tag{39}$$

where $\nu,\nu' = 1,\cdots,m$. Given the above information, we utilized the *fkf* function in the *FKF* package (version 0.2.3) to implement the Kalman filtering and smoothing

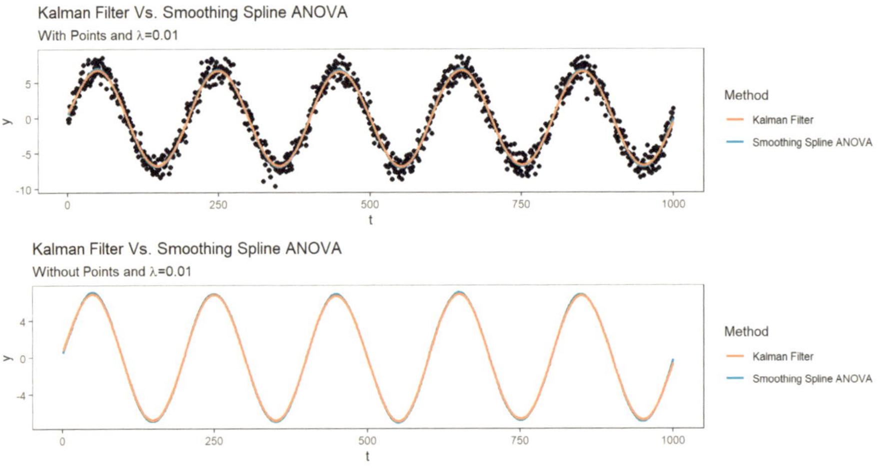

Figure 1.
Comparison between the SS-ANOVA model fit and the model fit with the Kalman filter given $\lambda = 0.01$ on simulated data.

algorithms for the SS-ANOVA model. The details of iteration procedures are available in the help document of *fkf*, which is similar to the procedures described in Section 3. **Figure 1** shows the similarity between the SS-ANOVA model fit with the generic algorithm from the *gss* package and the model fit based on the state space representations in (35) and (36).

6. Conclusions

In this chapter, we have introduced the theoretical foundation (e.g., representer theorem) and estimation algorithms of SS-ANOVA models. Given tensor product operations, the SS-ANOVA models can handle the multivariate data and study the main and interaction effects in the corresponding subspaces via functional ANOVA. The estimation algorithms of SS-ANOVA models need $O(N^3)$ operations, which might be prohibitive computationally for analyzing super large data. Utilizing the Markov structure of SS-ANOVA models, the Kalman filter can be used to fit SS-ANOVA models when reparameterized into a state space formulation [11]. This state space representation reduces the computational complexity from $O(N^3)$ to $O(N)$ for the univariate case, allowing SS-ANOVA models to be applicable to big data applications. Additional research has been done to extend this representation to the multidimensional setting [12]. For the two-dimensional data with the dimensions of n_1 and n_2, the SS-ANOVA models can be fitted with the computational complexity of $O(n_1 n_2^3)$, where $N = n_1 n_2$. Furthermore, we provided a simulated example to compare estimates from the state space representation and the estimates from the SS-ANOVA model for the univariate case.

Acknowledgements

Sun's research was supported by the 18th Mile TRIF Funding from the University of Arizona and the NIH grant U19AG065169. LaFleur's research was supported by the NIH grant U19AG065169.

Conflict of interest

The authors declare no conflict of interest.

Author details

Joel Parker[1,2], Yifan Zhang[1], Bonnie J. Lafleur[2,3] and Xiaoxiao Sun[1*]

1 Department of Epidemiology and Biostatistics, The University of Arizona, Tucson, Arizona, USA

2 BIO5 Institute, The University of Arizona, Tucson, Arizona, USA

3 R. Ken Coit College of Pharmacy, The University of Arizona, Tucson, Arizona, USA

*Address all correspondence to: xiaosun@arizona.edu

References

[1] Brinkman ND. Ethanol fuel-single-cylinder engine study of efficiency and exhaust emissions. SAE Transactions. 1981;**90**:1410-1424

[2] Gu C. Smoothing Spline ANOVA Models. New York, NY: Springer; 2013

[3] Wahba G. Spline Models for Observational Data. Philadelphia, PA: Society for Industrial and Applied Mathematics; 1990

[4] Kim YJ, Gu C. Smoothing spline Gaussian regression: More scalable computation via efficient approximation. Journal of the Royal Statistical Society: Series B (Statistical Methodology). 2004;**66**(2):337-356

[5] Ma P, Huang JZ, Zhang N. Efficient computation of smoothing splines via adaptive basis sampling. Biometrika. 2015;**102**(3):631-645

[6] Helwig NE, Ma P. Smoothing spline ANOVA for super-large samples: Scalable computation via rounding parameters. Statistics and Its Interface. 2016;**9**:433-444

[7] Sun X, Zhong W, Ma P. An asymptotic and empirical smoothing parameters selection method for smoothing spline ANOVA models in large samples. Biometrika. 2021;**108**(1): 149-166

[8] Kalman RE. A new approach to linear filtering and prediction problems. Journal of Basic Engineering. 1960;**82**(1): 35-45

[9] Moreno VM, Pigazo A. Kalman Filter: Recent Advances and Applications. London, UK: IntechOpen; 2009

[10] Shumway RH. Longitudinal data with serial correlation: A state-space approach. Journal of Statistical Planning and Inference. 1995;**47**(3):395-397

[11] Wecker WE, Ansley CF. The signal extraction approach to nonlinear regression and spline smoothing. Journal of the American Statistical Association. 1983;**78**(381):81-89

[12] Qin L, Guo W. State space representation for smoothing spline ANOVA models. Journal of Computational and Graphical Statistics. 2006;**15**(4):830-847

[13] Koopman SJ, Durbin J. Fast filtering and smoothing for multivariate state space models. Journal of Time Series Analysis. 2000;**21**(3):281-296

[14] Wang Y. Smoothing Splines: Methods and Applications. Boca Raton, FL: CRC Press; 2011

[15] Yang Y. Model selection for nonparametric regression. Statistica Sinica. 1999;**9**:475-499

[16] Abramowitz M, Stegun IA. Handbook of Mathematical Functions with Formulas, Graphs, and Mathematical Tables. Washington, DC: U.S. Govt. Print. Off; 1964

[17] Wahba G. A comparison of GCV and GML for choosing the smoothing parameter in the generalized spline smoothing problem. The Annals of Statistics. 1985;**13**:1378-1402

[18] Xiang D, Wahba G. A generalized approximate cross validation for smoothing splines with non-Gaussian data. Statistica Sinica. 1996:675-692

[19] Graven P, Wahba G. Smoothing noisy data with spline function:

Estimating the correct degree of smoothing by the method of Generalized Cross-Validaton. Numerische Mathematik. 1979;**31**:377-403

[20] De Jong P. The likelihood for a state space model. Biometrika. 1988;**75**(1): 165-169

[21] Douc R, Moulines E, Stoffer D. Nonlinear Time Series: Theory, Methods and Applications with R Examples. Boca Raton, FL: CRC Press; 2014

[22] Durbin J, Koopman SJ. A Simple and Efficient Simulation Smoother for State Space Time Series Analysis. Biometrika, 2002;**89**(3);603-615

[23] Wahba G. Improper priors, spline smoothing and the problem of guarding against model errors in regression. Journal of the Royal Statistical Society: Series B (Methodological). 1978;**40**(3): 364-372

[24] Gu C. Smoothing Spline ANOVA Models: R Package gss. Journal of Statistical Software. 2014;**58**(5):1-25

Chapter 5

Sequential Mini-Batch Noise Covariance Estimator

*Hee-Seung Kim, Lingyi Zhang, Adam Bienkowski,
Krishna R. Pattipati, David Sidoti, Yaakov Bar-Shalom
and David L. Kleinman*

Abstract

Noise covariance estimation in an adaptive Kalman filter is a problem of significant practical interest in a wide array of industrial applications. Reliable algorithms for their estimation are scarce, and the necessary and sufficient conditions for identifiability of the covariances were in dispute until very recently. This chapter presents the necessary and sufficient conditions for the identifiability of noise covariances, and then develops sequential mini-batch stochastic optimization algorithms for estimating them. The optimization criterion involves the minimization of the sum of the normalized temporal cross-correlations of the innovations; this is based on the property that the innovations of an optimal Kalman filter are uncorrelated over time. Our approach enforces the structural constraints on noise covariances and ensures the symmetry and positive definiteness of the estimated covariance matrices. Our approach is applicable to non-stationary and multiple model systems, where the noise covariances can occasionally jump up or down by an unknown level. The validation of the proposed method on several test cases demonstrates its computational efficiency and accuracy.

Keywords: Adaptive Kalman filtering, Minimal polynomial, Noise covariance estimation, Stochastic gradient descent (SGD), Mini-batch SGD

1. Introduction

This chapter addresses the following learning problem: Given a vector time series and a library of models for the time evolution of the data (e.g., a Wiener process, a white noise acceleration model, also called nearly constant velocity model, or a white noise jerk model, also called nearly constant acceleration model), find suitable process and measurement noise covariances and select the best dynamic model for the time series. This problem is of considerable interest in a number of applications, such as fault diagnosis, robotics, signal processing, navigation, and target tracking, to name a few [1, 2].

The Kalman filter (KF) [3] is the optimal minimum mean square error (MMSE) state estimator for linear systems with mutually uncorrelated Gaussian white process and measurement noises, and is the best linear state estimator when the noises are non-Gaussian with known covariances. However, the noise covariances are unknown or only partially known in many practical applications.

We derived the necessary and sufficient conditions for the identifiability of unknown noise covariances, and presented a batch optimization algorithm for their estimation using the sum of the normalized temporal cross-correlations of the innovation sequence as the optimization criterion [4]. The motivation for this optimization metric stems from the fact that the innovations of an optimal Kalman filter are white, meaning that they are uncorrelated over time [2]. In [5], we proposed a sequential mini-batch stochastic gradient descent (SGD) algorithm that required multiple passes through the measurements for estimating noise covariances. We also presented its applicability to non-stationary systems by detecting changes in noise covariances. In this chapter, we present a practical single-pass stochastic gradient descent algorithm for noise covariance estimation in non-stationary systems. Extensions to multiple models where the system behavior can stem from a member of a known subset of models are discussed in [6].

1.1 Prior work

The key to noise covariance estimation is an expression for the covariance of the state estimation error and of the innovations of any stable, but not necessarily optimal, filter as a function of noise covariances. This expression serves as a foundational building block for the correlation-based methods for noise covariance estimation. Pioneering contributions using this approach were made by [7–9]. Sarkka and Nummenmaa [10] proposed a recursive noise-adaptive Kalman filter for linear state space models using variational Bayesian approximations. However, the variational methods generally require tuning hyper-parameters to converge to the correct covariance parameters and these algorithms often converge to local minima.

In [5], we presented a computationally efficient and accurate sequential estimation algorithm that is a major improvement over the batch estimation algorithm in [4]. The novelties of this algorithm stem from its sequential nature and the use of mini-batches, adaptive step size rules and dynamic thresholds for convergence in the stochastic gradient descent (SGD) algorithm. The innovation cross-correlations are obtained by a sequential fading memory filter. We applied a change-point detection algorithm described in [11] to extract the change points in noise covariances for non-stationary systems.

This chapter seeks to develop a streaming algorithm that reads measurements exactly once, thus making it real-time and practical. The only caveat is that the changes in noise covariances are assumed to occasionally jump up or down by an unknown magnitude. Extensions of this algorithm to a multiple model setting may be found in [6].

1.2 Organization of the chapter

The organisation of the chapter is as follows. Section 2 presents the mathematical formulation of the sequential mini-batch gradient descent algorithm for estimating the unknown noise covariances. In this section, we also present an overview of our approach based on a fading memory filter-based innovation correlation estimation, and an accelerated SGD update of the Kalman gain. In Section 3, we show that our single-pass method can track unknown noise covariances in non-stationary systems. Lastly, we conclude the chapter with a brief summary of the contributions in Section 4.

2. Sequential mini-batch SGD method for estimating process and measurement noise covariances

Consider a discrete-time linear dynamic system

$$x(k+1) = Fx(k) + \Gamma v(k) \tag{1}$$

$$z(k) = Hx(k) + w(k) \tag{2}$$

where $x(k)$ is the n_x-dimensional state vector, $v(k)$ is the sequence of zero-mean white Gaussian process noise with unknown process noise covariance $Q(k)$ in the plant equation. The measurement equation, $z(k)$ with n_z-dimensional vector, is given in (2). Here, $w(k)$ is the sequence of zero-mean white Gaussian measurement noise with unknown measurement noise covariance $R(k)$. In the system, F and Γ are the $n_x \times n_x$ state transition matrix and the noise gain matrix, respectively, and H is the $n_z \times n_x$ measurement matrix. Here, the two noise sequences and the initial state error are assumed to be mutually uncorrelated. We assume that noise covariances $Q(k)$ and $R(k)$ are piecewise constant such that the filter reaches a steady-state between any two jumps of unknown magnitude.

Given $Q(k)$ and $R(k)$, the Kalman filter involves the consecutive processes of prediction and update given by [2, 3].

$$\hat{x}(k+1|k) = F\hat{x}(k|k) \tag{3}$$

$$\nu(k+1) = z(k+1) - H\hat{x}(k+1|k) \tag{4}$$

$$\hat{x}(k+1|k+1) = \hat{x}(k+1|k) + W(k+1)\nu(k+1) \tag{5}$$

$$P(k+1|k) = FP(k|k)F' + \Gamma Q(k)\Gamma' \tag{6}$$

$$S(k+1) = HP(K+1|k)H' + R(k) \tag{7}$$

$$W(k+1) = P(k+1|k)H'S(k+1)^{-1} \tag{8}$$

$$\begin{aligned} P(k+1|k+1) = {} & (I_{n_x} - W(k+1)H)P(k+1|k)(I_{n_x} - W(k+1)H)' \\ & + W(k+1)R(k)W(k+1)' \end{aligned} \tag{9}$$

The Kalman filter predicts the next state estimate at time index $(k+1)$, given the observations up to time index k in (3) and the concomitant predicted state estimation error covariance in (6), using system dynamics, the updated state error covariance $P(k|k)$ at time index k and the process noise covariance, $Q(k)$. The updated state estimate at time $(k+1)$ in (5) incorporates the measurement at time $(k+1)$ via the Kalman gain matrix in (8), which depends on the innovation covariance $S(k+1)$ (which in turn depends on the measurement noise covariance $R(k)$, and the predicted state error covariance $P(k+1|k)$). The updated state error covariance $P(k+1|k+1)$ is computed via (9); this is the Joseph form, which is less sensitive to round-off error because it guarantees that the updated state covariance matrix will remain positive definite.

2.1 Identifiability conditions for estimating Q and R

The necessary and sufficient conditions for identifiability of the covariances in adaptive Kalman filters were in dispute until very recently [4, 7–9, 12]. When Q and R are unknown, consider the innovations corresponding to a stable, suboptimal closed-loop filter matrix $\overline{F} = F(I_{n_x} - WH)$ given by [4, 13].

$$\begin{aligned}\nu(k) &= H\overline{F}^{m}[x(k-m) - \hat{x}(k-m|k-m-1)] \\ &\quad + \left\{ H\sum_{j=0}^{m-1} \overline{F}^{m-1-j}[\Gamma v(k-m+j) - FWw(k-m+j)] \right\} + w(k)\end{aligned} \tag{10}$$

Given the innovation sequence (10), a weighted sum of innovations, $\xi(k)$, can be computed as

$$\xi(k) = \sum_{i=0}^{m} a_i \nu(k-i) \tag{11}$$

where the weights are the coefficients of the minimal polynomial of the closed-loop filter matrix $\overline{F}$, $\sum_{i=0}^{m} a_i \overline{F}^{m-i} = 0, a_0 = 1$. It is easy to see that $\xi(k)$ is the sum of two moving average processes driven by the process noise and measurement noise, respectively, given by [4].

$$\xi(k) = \sum_{l=1}^{m} \mathcal{B}_l v(k-l) + \sum_{l=0}^{m} \mathcal{G}_l w(k-l) \tag{12}$$

Here, $\mathcal{B}_l$ and $\mathcal{G}_l$ are given by

$$\mathcal{B}_l = H\left(\sum_{i=0}^{l-1} a_i \overline{F}^{l-i-1}\right)\Gamma \tag{13}$$

$$\mathcal{G}_l = \left[a_l I_{n_z} - H\left(\sum_{i=0}^{l-1} a_i \overline{F}^{l-i-1}\right)FW\right]; \mathcal{G}_0 = I_{n_z} \tag{14}$$

Then, the cross-covariance between $\xi(k)$ and $\xi(k-j)$, $\mathcal{L}_j$, can be obtained as

$$\mathcal{L}_j = E[\xi(k)\xi(k-j)'] = \sum_{i=j+1}^{m} \mathcal{B}_i Q \mathcal{B}_{i-j}' + \sum_{i=j}^{m} \mathcal{G}_i R \mathcal{G}_{i-j}' \tag{15}$$

The noise covariance matrices $Q = [q_{ij}]$ of dimension $n_v \times n_v$ and $R = [r_{ij}]$ of dimension $n_z \times n_z$ are positive definite and symmetric. By converting the noise covariance matrices and the $\mathcal{L}_j$ matrices to vectors, Zhang et al. [4] show that they are related by the noise covariance identifiability matrix $\mathcal{I}$ given by

$$\mathcal{I}\begin{bmatrix} \text{vec}(Q) \\ \text{vec}(R) \end{bmatrix} = \begin{bmatrix} \mathcal{L}_0 \\ \mathcal{L}_1 \\ \vdots \\ \mathcal{L}_m \end{bmatrix} \tag{16}$$

As shown in [4], if the matrix $\mathcal{I}$ has full rank, then the unknown noise covariance matrices, Q and R, are identifiable. Direct solution of linear equations in (16) for Q and R is highly ill-conditioned and is prone to numerical errors.

2.2 Recursive fading memory-based innovation correlation estimation

We compute the sample correlation matrix $\hat{C}^{k}_{seq}(i)$ at sample k for time lag i as a weighted combination of the correlation matrix $\hat{C}^{k-1}_{seq}(i)$ at the previous sample $(k-1)$ and time lag i, and the samples of innovations $\nu(k-i)$ and $\nu(k)$. The tuning parameter λ, a positive constant between 0 and 1, is the weight associated with the previous sample correlation matrix. The current M sample correlation matrices at time k are used as the initial values for the next pairs of samples for the recursive computation. Let us define the number of measurement samples as N. Then,

$$\hat{C}^{k}_{seq}(i) = (1-\lambda)\nu(k-i)\nu(k)' + \lambda \hat{C}^{k-1}_{seq}(i), \tag{17}$$

$$\hat{C}^{0}_{seq}(i) = 0, i = 0,1,2,\cdots,M-1; k = M,\cdots,N \tag{18}$$

2.3 Objective function and the gradient

The ensemble cross-correlation of a steady-state suboptimal Kalman filter is related to the closed-loop filter matrix $\overline{F} = F(I_{n_x} - WH)$, the matrix F, the measurement matrix H, the steady-state predicted covariance matrix $\overline{P}$, steady-state filter gain W and the steady-state innovation covariance, $C(0)$ via [8, 9].

$$C(i) = E\left[\nu(k)\nu(k-i)'\right] = H\overline{F}^{i-1}F\left[\overline{P}H' - WC(0)\right] \tag{19}$$

To avoid the scaling effects of measurements, the objective function Ψ formulated in [4] involves a minimization of the sum of normalized $C(i)$ with respect to the corresponding diagonal elements of $C(0)$ for $i > 0$. Formally, we can define the objective function Ψ to be minimized with respect to W as

$$\Psi = \frac{1}{2} tr\left\{\sum_{i=1}^{M-1} [diag(C(0))]^{-\frac{1}{2}} C(i)' [diag(C(0))]^{-1} C(i) [diag(C(0))]^{-\frac{1}{2}}\right\} \tag{20}$$

where $diag(C)$ denotes the diagonal matrix of C. We can rewrite the objective function by substituting (20) into (19) as

$$\Psi = \frac{1}{2} tr\left\{\sum_{i=1}^{M-1} \phi(i) X \varphi X'\right\} \tag{21}$$

where

$$\phi(i) = \left[H\overline{F}^{i-1}F\right]' \varphi \left[H\overline{F}^{i-1}F\right] \tag{22}$$

$$X = \overline{P}H' - WC(0) \tag{23}$$

$$\varphi = [diag(C(0))]^{-1} \tag{24}$$

The gradient of objective function, $\nabla_W \Psi$, can be computed as [4].

$$\nabla_W \Psi = -\sum_{i=1}^{M-1} \left[H\overline{F}^{i-1}F\right]' \varphi C(i) \varphi C(0) - F'ZFX - \sum_{l=0}^{i-2} \left[C(l+1) \varphi C(i)' \varphi H \overline{F}^{i-l-2}\right]' \tag{25}$$

where

$$Z = \overline{F}' Z \overline{F} + \frac{1}{2} \sum_{i=1}^{M-1} \left(H\overline{F}^{i-1}F\right)' \varphi C(i) \varphi H + \left(\left(H\overline{F}^{i-1}F\right)' \varphi C(i) \varphi H\right)' \tag{26}$$

The Z term in (26) is computed by a Lyapunov equation; it is often small and can be neglected in (25) for computational efficiency.

In computing the objective function and the gradient, we replace $C(i)$ by their sample estimates, $\hat{C}_{seq}(i)$. With this replacement, the noise covariance estimation becomes a data-dependent stochastic optimization/learning problem.

2.4 Estimation of Q and R

2.4.1 Estimation of R

We define $\mu(k)$ as the post-fit residual sequence of the Kalman filter, which is related to the innovations $\nu(k)$ via

$$\mu(k) = z(k) - H\hat{x}(k|k) = (I_{n_z} - HW)\nu(k); k = 1,2,\cdots,N \tag{27}$$

From the joint covariance of the innovation sequence $\nu(k)$ and the post-fit residual sequence $\mu(k)$, and the Schur determinant identity [14, 15], one can show that [4].

$$G = E\left[\mu(k)\mu(k)'\right] = RS^{-1}R \tag{28}$$

where S is the innovation covariance. Knowing the sampled estimates of G and S=$\hat{C}_{seq}(0)$, the measurement noise covariance R is estimated. Because (28) can be interpreted as a continuous-time algebraic Riccati equation or as a simultaneous diagonalization problem in linear algebra [15], the measurement noise covariance R can be estimated by solving a continuous-time Riccati equation as in [4, 16] or by solving the simultaneous diagonalization problem via Cholesky decomposition and eigen decomposition.

2.4.2 Estimation of Q

Since the process noise covariance Q and the steady-state updated covariance P are generally coupled, Q and P can be obtained via a Gauss–Seidel type iterative computation given the estimated R. Wiener process is an exception where an explicit non-iterative solution $Q = WSW'$ is possible [4]. Let t and l denote the indices of iteration starting with $t = 0$ and $l = 0$. The initial steady-state updated covariance, P^0, can be computed as the solution of the Lyapunov equation given by

$$P^{(0)} = \tilde{F}P^{(0)}\tilde{F}' + WRW' + (I_{n_x} - WH)\Gamma Q^{(t)}\Gamma'(I_{n_x} - WH)'; Q^{(0)} = WSW' \quad (29)$$

where $\tilde{F} = (I_{n_x} - WH)F$. We iteratively update P as in (30) until convergence

$$P^{(l+1)} = \left[\left(FP^{(l)}F' + \Gamma Q^{(t)}\Gamma'\right)^{-1} + H'R^{-1}H\right]^{-1} \quad (30)$$

Given the converged P, Q will be updated in the t-loop until the estimate of Q converges.

$$Q^{(t+1)} = \Gamma^{\dagger}\left[(P + WSW' - FPF')^{(t+1)} + \lambda_Q I_{n_x}\right](\Gamma')^{\dagger} \quad (31)$$

where λ_Q is a regularization parameter used for ill-conditioned estimation problems.

2.5 Updating the gain W sequentially

The estimation algorithm sequentially computes the M sample covariance matrices at every measurement sample k as in (17). Let B be the mini-batch size for updating the Kalman filter gain W in the SGD. Our proposed method updates the gain W when the sample index k is divisible by the mini-batch size B. When compared to the batch estimation algorithm, the sequential mini-batch SGD algorithm allows more opportunities to converge to a better local minimum of (20) by frequently updating the filter the gain [5]. The generic form of the gain update is given by

$$W^{(r+1)} = W^{(r)} - \alpha^{(r)}\nabla_{W^{(r)}}\Psi \quad (32)$$

where r is the updating index starting with $r = 0$. In our previous research [5], we explored the performance of accelerated SGD methods (e.g., bold driver [17], constant, subgradient [18], RMSProp [19], Adam [20], Adadelta [21]) for updating adaptive step size $\alpha^{(r)}$ in (32). The root mean square propagation (RMSProp) method is applied for the estimation procedure in this chapter. The RMSProp keeps track of the moving average of the squared incremental gradients for each gain element by adapting the step size element-wise as in the following.

$$\tau_{r,ij} = \gamma\tau_{r-1,ij} + (1-\gamma)\left[\left(\nabla_{W^{(r)}}\Psi\right)_{ij}\right]^2; \tau_0 = 0 \quad (33)$$

$$\alpha_{ij}^{(r)} = \frac{\alpha^{(0)}}{\sqrt{\tau_{r,ij} + \varepsilon}} \quad (34)$$

Here, $\gamma = 0.9$ is the default value and $\varepsilon = 10^{-8}$ to prevent division by zero.

The pseudocode for the sequential mini-batch SGD estimation algorithm for a non-stationary system is included as Algorithm 1.

Algorithm 1 Pseudocode of sequential mini-batch SGD algorithm.

input: W^0, Q^0, R^0, α^0, B ▷ W^0: initial gain, Q^0: initial Q, R^0: initial R, α^0: initial step size, B: batch size.
r = 0 {▷ Initialize the updating index r}.
for k = 1 to N **do** {▷ N: Number of samples}.
  compute innovation correlations $\nu(k)$.
  if $k > N_b + M$ **then** {▷ N_b: Number of burn-in samples}.
    compute $\hat{C}^k_{seq}(i)$, i = 0,1,2,...M-1.
    if $Mod(k, B) = 0$ **then.**
      compute the objective function Ψ.
      compute the gradient $\nabla_W \Psi$.
      update the step size $\alpha^{(r)}$.
      update the gain $W_{ij}^{(r+1)} = W_{ij}^{(r)} - \alpha_{ij}^{(r)} [\nabla_{W^{(r)}} \Psi)]_{ij}$.
      update $R^{(r+1)}$ and $Q^{(r+1)}$.
      r = r + 1.
    end if.
  end if.
end for.

3. Numerical examples

In [5], we provided the evidence that the multi-pass sequential mini-batch stochastic gradient descent (SGD) algorithms improve the computational efficiency of the batch estimation algorithm via a number of test cases used in [2, 7–9, 12], and also showed their applicability to non-stationary systems when coupled with a change-point detection algorithm [11]. In [22], we proposed a single-pass sequential mini-batch SGD estimation algorithm by accessing measurements exactly once for non-stationary systems by modifying the example used in [12] to periodically change the process and measurement noise covariances.

In this section, we illustrate the utility of our proposed single-pass sequential mini-batch SGD estimation algorithm by applying it to general diverse examples involving detectable (but not completely observable) systems, non-stationary systems and a bearings-only tracking problem.

For the non-stationary systems, we assumed the process and measurement noise covariances occasionally change by an unknown level. Here "occasionally" implies the jumps are infrequent enough that the Kalman filter is in the steady-state prior to a jump in the noise covariance. We define the number of subgroups in which the noise covariances are not changing as N_{sg}. Given the number of observation samples, N, each subgroup has constant noise covariances with N/N_{sg} samples. In this section, we consider two non-stationary scenarios for tracking time-varying Q and R with $N_{sg} = 5$ subgroups. We also consider the bearings-only tracking problem where Q changes continuously.

Note that the number of "burn-in" samples and the number of lags are $N_b = 50$ and $M = 5$, respectively in the estimation procedure. The root mean square propagation (RMSProp) method is applied to update the filter gain. All Monte Carlo simulations were run using a computer with an Intel Core i7-8665U processor and 16 GB of RAM.

We used the averaged normalized innovation squared (NIS) metric [2] to measure the consistency of the proposed algorithm.

$$\bar{\varepsilon}_{\nu}(k) = \frac{1}{N_{mc}} \sum_{i=1}^{N_{mc}} \nu(k)' S^{-1} \nu(k) \tag{35}$$

where N_{mc} is the number of Monte Carlo runs. The root mean square error (RMSE) in resultant position and velocity is computed using

$$RMSE(k) = \sqrt{\frac{1}{N_{mc}} \sum_{i=1}^{N_{mc}} \left(x(k)_{true} - \hat{x}(k)\right)^2} \tag{36}$$

3.1 Case 1: A detectable (but not completely observable) system that satisfies the identifiability conditions

Mehra [8] stated, without proof, that a necessary and sufficient condition for noise covariance estimation is that the system satisfies the observability property. This example, due to Odelson et al. [7], demonstrates that this condition is not necessary. The example does satisfy the full column rank condition for the identifiability matrix in (16).

Odelson et al. [7] proposed a noise covariance estimation method based on the autocovariance least-squares formulation by using the Kronecker operator δ. This method computes the covariances from the residuals of the state estimation. Note that the incompletely observable (but detectable [1]) system used in [7] is described by

$$F = \begin{bmatrix} 0.1 & 0 \\ 0 & 0.2 \end{bmatrix}, \; H = [1 \;\; 0], \; \Gamma = \begin{bmatrix} 1 \\ 2 \end{bmatrix} \tag{37}$$

where F is the non-singular transition matrix, H is the constant output matrix, and Γ is the constant input matrix in (1) and (2). Note that this system is a hypothetical numerical example. The process noise $v(k)$ and the measurement noise $w(k)$ are supposed to be uncorrelated Gaussian white noise sequences with zero-mean and covariances as in the following

$$E\left[v(k)v(j)'\right] = Q\delta_{kj} \tag{38}$$

$$E\left[w(k)w(j)'\right] = R\delta_{kj} \tag{39}$$

In this scenario, the true R values for the five subgroups are [0.30, 0.81, 0.49, 0.72, 0.42], and the true Q values for the five subgroups are [0.16, 0.49, 0.25, 0.36, 0.20]. The values are changed every 10,000 samples. **Table 1** shows the results of 100 Monte Carlo simulations based on the single-pass SGD algorithm in estimating Q and R. As can be seen, the estimated Q and R are close to their corresponding true values. In this

[1] The pair (F, H) in the system should be detectable in order for the continuous-time algebraic Riccati equation to have at least one positive semidefinite solution and in this case at least one solution results in a marginally stable steady-state KF [23, 24].

Subgroup index	R			Q			$\overline{P}_{11}$			$\overline{P}_{22}$			W_{11}			W_{21}		
	Truth	**Mean**	**RMSE**	**Truth**	**Mean**	**RMSE**	**Truth**	**Mean**	**RMSE**	**Truth**	**Mean**	**RMSE**	**Truth**	**Mean**	**RMSE**	**Truth**	**Mean**	**RMSE**
1st	0.30	0.31	0.06	0.16	0.15	0.04	0.16	0.15	0.04	0.66	0.62	0.18	0.35	0.33	0.03	0.70	0.67	0.03
2nd	0.81	0.86		0.49	0.42		0.49	0.43		2.01	1.74		0.38	0.34		0.76	0.67	
3rd	0.49	0.49		0.25	0.24		0.25	0.24		1.03	0.97		0.34	0.34		0.68	0.67	
4th	0.72	0.71		0.36	0.35		0.36	0.35		1.48	1.44		0.33	0.34		0.67	0.67	
5th	0.42	0.41		0.20	0.20		0.20	0.21		0.83	0.84		0.33	0.34		0.66	0.67	

Table 1.
Single-pass SGD estimation for Case 1 (100 MC Runs; 50,000 samples; M = 5; RMSProp; Batch size = 64).

scenario, the single-pass SGD estimation method has a speedup factor of 31 over the batch and multiple-pass SGD estimation methods (not shown).

Figure 1 demonstrates that the sequential mini-batch gradient descent algorithm can track Q and R correctly. Here, the trajectories of Q and R estimates are smoothed by a simple first order fading memory filter with a smoothing weight of 0.7. **Figure 1e** shows the averaged NIS of SGD (RMSProp; batch size of 64) method with the 95% probability region [0.74, 1.30], and shows that the SGD-based Kalman filter is consistent. The only place at which the NIS values are large are immediately after the jump in the noise variances. This is because adaptation requires a few samples.

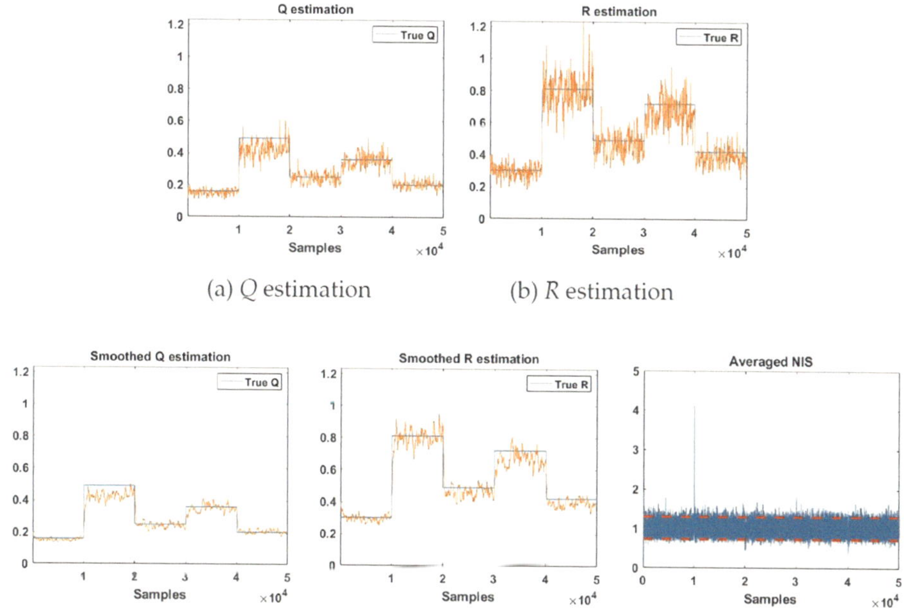

(a) Q estimation (b) R estimation

(c) Smoothed Q estimation (d) Smoothed R estimation (e) Averaged NIS with 95% probability region [0.74, 1.30]

Figure 1.
Trajectories of Q and R estimates without signal smoothing and with a smoothing weight of 0.7 for Case 1.

3.2 Case 2: a five-state inertial navigation system with diagonal Q and R

For estimating the unknown noise covariance parameters and the optimal Kalman filter gain for part of an inertial navigation system (INS), Mehra [8] proposed an iterative innovation correlations-based method starting from an arbitrary initial stabilizing gain. Inertial navigation [25] involves tracking the position and orientation of an object relative to a known starting orientation and velocity and it uses measurements provided by accelerometers and gyroscopes. These systems have found universal use in military and commercial applications [26].

Since the earth is not flat, the inertial navigation systems need to keep tilting the platform (with respect to inertial space) to keep the axes of the accelerometers horizontal. Here, small error sources that drive the Schuler-loop cause the navigation errors, and these errors are" damped" by making use of external velocity measurements, such as are furnished by a Doppler radar [27, 28].

In this problem, Mehra [8] used a system based on the damped Schuler-loop error propagation forced by exponentially correlated as well as white noise input. The system matrices for this navigation system are given by

$$F = \begin{bmatrix} 0.75 & -1.74 & -0.3 & 0 & -0.15 \\ 0.09 & 0.91 & -0.0015 & 0 & -0.008 \\ 0 & 0 & 0.95 & 0 & 0 \\ 0 & 0 & 0 & 0.55 & 0 \\ 0 & 0 & 0 & 0 & 0.905 \end{bmatrix}; H = \begin{bmatrix} 1 & 0 & 0 & 0 & 1 \\ 0 & 1 & 0 & 1 & 0 \end{bmatrix};$$
$$\Gamma = \begin{bmatrix} 0 & 0 & 0 \\ 0 & 0 & 0 \\ 24.64 & 0 & 0 \\ 0 & 0.835 & 0 \\ 0 & 0 & 1.83 \end{bmatrix} \quad (40)$$

where the system is discretized using a time step of 0.1 seconds. In this five-state system, the first two states represent the a velocity damping term and velocity error, respectively, and the other three states model the correlated noise processes. The noise corresponding to states 3 and 5 impacts both the velocity error and the velocity damping term; the fourth state impacts the sensor error in state 2 only.

In this problem, the true values corresponding to each subgroup with $N_{sg} = 5$ subgroups and $N = 100{,}000$ samples are as in (41). Each parameter of Q and R changes every 20,000 samples.

$$\begin{bmatrix} R_{11} \\ R_{22} \end{bmatrix} = \begin{bmatrix} [0.25,\ 0.56,\ 0.64,\ 0.42,\ 0.36] \\ [0.25,\ 0.25,\ 0.49,\ 0.16,\ 0.04] \end{bmatrix}; \begin{bmatrix} Q_{11} \\ Q_{22} \\ Q_{33} \end{bmatrix} = \begin{bmatrix} [0.25,\ 0.64,\ 0.49,\ 0.25,\ 0.49] \\ [0.25,\ 0.36,\ 0.56,\ 0.16,\ 0.04] \\ [0.36,\ 0.49,\ 0.64,\ 0.25,\ 0.09] \end{bmatrix} \quad (41)$$

Table 2 shows the results of 100 Monte Carlo simulations for estimating the noise parameters using RMSProp update. The estimated parameters are close to the corresponding true values. Given 100,000 samples, the proposed method with a batch-size of 64 requires 1,891 seconds for 100 Monte Carlo simulations, i.e., 18.91 seconds per run. The batch and multi-pass SGD estimation methods need more than 3,000 seconds for a single MC run (not shown); the single-pass SGD algorithm has a speedup factor of 158.6.

Figure 2 shows the trajectories of the estimated Q and R with a signal smoothing with a smoothing weight of 0.7. For this example, it is known that accurate estimation of R_{11} is hard as shown in **Figure 2d**. The reason is that R_{11} is dominated by the state uncertainty, i.e., the measurement noise is "buried" in a much larger innovation [4]. In spite of the difficulty in estimating R_{11}, the filter is consistent as measured by NIS as shown in **Figure 2f**.

(a) R, Q and $\overline{P}$ Estimates for Case 2

Subgroup index	R_{11}		RMSE	R_{22}		RMSE	Q_{11}		RMSE	Q_{22}		RMSE	Q_{33}		RMSE
	Truth	Mean		Truth	Mean		Truth	Mean		Truth	Mean		Truth	Mean	
1st	0.25	0.23	0.34	0.25	0.24	0.04	0.25	0.24	0.05	0.25	0.23	0.05	0.36	0.35	0.06
2nd	0.56	0.52		0.25	0.25		0.64	0.62		0.36	0.31		0.49	0.44	
3rd	0.64	0.85		0.49	0.50		0.49	0.50		0.56	0.52		0.64	0.71	
4th	0.42	0.44		0.16	0.16		0.25	0.26		0.16	0.16		0.25	0.25	
5th	0.36	0.30		0.04	0.04		0.49	0.48		0.04	0.04		0.09	0.08	

Subgroup index	$\overline{P}_{11}$		RMSE	$\overline{P}_{22}$		RMSE	$\overline{P}_{33}$		RMSE	$\overline{P}_{44}$		RMSE	$\overline{P}_{55}$		RMSE
	Truth	Mean		Truth	Mean		Truth	Mean		Truth	Mean		Truth	Mean	
1st	19.46	18.84	4.12	0.33	0.31	0.05	306.26	297.69	73.27	0.23	0.22	0.05	4.02	3.86	0.71
2nd	43.76	42.11		0.43	0.40		766.59	744.98		0.34	0.30		5.45	4.99	
3rd	37.81	40.15		0.66	0.67		600.91	621.04		0.52	0.49		7.33	8.00	
4th	18.49	18.96		0.22	0.22		303.75	312.24		0.15	0.15		2.78	2.78	
5th	29.31	28.55		0.07	0.07		574.37	562.90		0.04	0.04		0.97	0.93	

(b) W Estimates for Case 2

Subgroup index	W_{11}		RMSE	W_{21}		RMSE	W_{31}		RMSE	W_{41}		RMSE	W_{51}		RMSE
	Truth	Mean		Truth	Mean		Truth	Mean		Truth	Mean		Truth	Mean	
1st	0.94	0.96	0.02	$2.78 \cdot 10^{-3}$	$-3.56 \cdot 10^{-3}$	0.01	−2.80	−2.75	0.04	$-2.22 \cdot 10^{-5}$	$6.50 \cdot 10^{-3}$	0.01	0.04	0.04	0.01
2nd	0.96	0.98		$3.22 \cdot 10^{-3}$	$-4.32 \cdot 10^{-3}$		−2.91	−2.88		$-8.76 \cdot 10^{-4}$	$8.37 \cdot 10^{-3}$		0.03	0.01	
3rd	0.94	0.95		$2.84 \cdot 10^{-3}$	$-3.84 \cdot 10^{-3}$		−2.79	−2.83		$-1.19 \cdot 10^{-4}$	$8.79 \cdot 10^{-3}$		0.04	0.03	
4th	0.94	0.95		$3.84 \cdot 10^{-3}$	$2.08 \cdot 10^{-4}$		−2.80	−2.84		$-6.80 \cdot 10^{-4}$	$3.89 \cdot 10^{-4}$		0.03	0.02	
5th	0.98	1.00		$3.50 \cdot 10^{-3}$	$-1.33 \cdot 10^{-4}$		−3.03	−3.00		$-1.09 \cdot 10^{-3}$	$2.10 \cdot 10^{-4}$		0.01	0.01	

Subgroup index	W_{12}		RMSE	W_{22}		RMSE	W_{32}		RMSE	W_{42}		RMSE	W_{52}		RMSE
	Truth	Mean		Truth	Mean		Truth	Mean		Truth	Mean		Truth	Mean	
1st	0.94	0.94	0.08	0.38	0.39	0.02	−1.63	−1.65	0.25	0.23	0.23	0.02	−0.94	−0.93	0.05
2nd	1.02	1.00		0.40	0.42		−1.65	−1.70		0.27	0.25		−1.02	−0.98	
3rd	0.86	0.96		0.36	0.37		−1.56	−1.68		0.26	0.25		−0.86	−0.93	
4th	0.99	0.98		0.39	0.39		−1.57	−1.69		0.23	0.23		−0.98	−0.99	
5th	1.19	1.06		0.44	0.46		−1.26	−1.77		0.20	0.22		−1.17	−1.11	

Table 2.
Single-pass SGD estimation for Case 2 (100 MC Runs, 100,000 samples; M = 5; RMSProp; Batch size = 64).

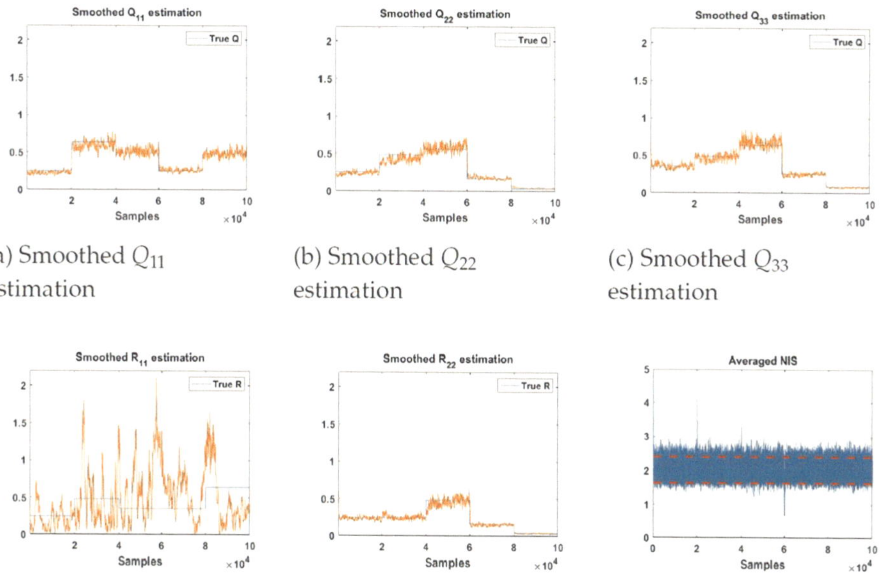

(a) Smoothed Q_{11} estimation (b) Smoothed Q_{22} estimation (c) Smoothed Q_{33} estimation

(d) Smoothed R_{11} estimation (e) Smoothed R_{22} estimation (f) Averaged NIS with 95% probability region [1.62, 2.41]

Figure 2.
Trajectories of Q and R estimates with a signal smoothing at smoothing weight = 0.7 for Case 2.

3.3 Case 3: Bearings-only tracking problem

In many practical situations, it is generally hard to get a closed-form solution for state estimation because the noise covariances are often unknown and the dynamics are nonlinear. Arasaratnam et al. [29] proposed a nonlinear filter using bearings-only measurements for estimating the position and velocity of a target in a high-dimensional state. This method is based on the measurements from a passive sensor that measures only the direction of arrival of a signal emitted by the target [2]. This so-called bearings-only tracking problem arises in a variety of practical applications, such as air traffic control, underwater sonar tracking and aircraft surveillance [2, 30, 31].

In this example, we consider a two-dimensional bearings-only tracking problem of a nearly-constant velocity target from a single moving observer used in [32]. The dynamics of the target (relative to the observer) are described by

$$x(k+1) = Fx(k) + \Gamma v(k) - U(k) \tag{42}$$

$$z(k) = h(x(k)) + w(k) \tag{43}$$

Formally, if the state vector of the target is $x^t(k) = \left[\zeta^t, \eta^t, \dot{\zeta}^t, \dot{\eta}^t\right]'$, and the state vector of the observer is $x^o(k) = \left[\zeta^o, \eta^o, \dot{\zeta}^o, \dot{\eta}^o\right]'$ for position and velocity along the ζ and η axes, $x(k) = x^t(k) - x^o(k)$ represents the relative state vector of the target with respect to the observer and the input vector $U(k) = x^o(k) - Fx^o(k-1)$; $w(k)$ is a zero-mean white Gaussian noise with variance σ_θ^2. The nonlinear measurement involves the bearing of the target from the observer's platform, given by $h(x(k)) = \tan^{-1}(\zeta/\eta)$. Here, Γ is the identity matrix with ones on the diagonal and zeros elsewhere.

The system matrices for this problem are given by

$$F = \begin{bmatrix} 1 & 0 & T & 0 \\ 0 & 1 & 0 & T \\ 0 & 0 & 1 & 0 \\ 0 & 0 & 0 & 1 \end{bmatrix}, \Gamma = \begin{bmatrix} 1 & 0 & 0 & 0 \\ 0 & 1 & 0 & 0 \\ 0 & 0 & 1 & 0 \\ 0 & 0 & 0 & 1 \end{bmatrix}, Q = \begin{bmatrix} T^3/3 & 0 & T^2/2 & 0 \\ 0 & T^3/3 & 0 & T^2/2 \\ T^2/2 & 0 & T & 0 \\ 0 & T^2/2 & 0 & T \end{bmatrix} \tilde{q}(k) \tag{44}$$

where the sampling interval, T is 1 second. The zero-mean white process noise intensity $\tilde{q}(k)$ is $\tilde{q}_0 = 9 \cdot 10^{-12}\, km^2/s^3$, except for the interval where it starts to increase rapidly to 1.5 $\cdot\tilde{q}_0$ around the sample index k = 480 and then decreases again rapidly around k = 960 as below:

$$\tilde{q}(k) = \begin{cases} \tilde{q}_0 + 0.25\tilde{q}_0(1 + \tanh(0.015(k - 480))), & k \leq 720 \\ \tilde{q}_0 + 0.25\tilde{q}_0(1 + \tanh(0.015(960 - k))), & \text{otherwise} \end{cases} \tag{45}$$

The linearized measurement matrix, $H(k)$, is the Jacobian of the measurement function given by

$$H_k = \frac{\partial h(x(k))}{\partial x(k)} = \begin{bmatrix} \frac{\eta(k)}{\zeta^2(k) + \eta^2(k)} & \frac{-\zeta(k)}{\zeta^2(k) + \eta^2(k)} & 0 & 0 \end{bmatrix} \tag{46}$$

A total of 1920 measurement samples were generated for this scenario. The observer moves straight with a speed of 5 *knots*, except for 480 seconds (between k = 480 and k = 960), where it turns with 2.4°/s as shown in **Figure 3** (these times are marked by cross sign).

For a fair comparison of the estimation algorithms, we initialized all filters with the same mean and covariance using the prior knowledge of the initial target range

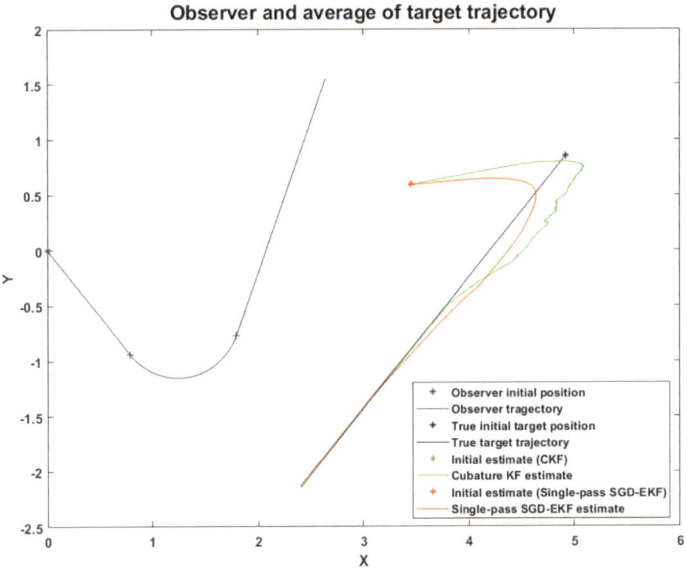

Figure 3.
Observer and target trajectory (100 MC runs).

and the initial bearing measurement [33, 34]. Here, the initial target range and the initial bearing measurement are generated as $\bar{r} \sim N(r, \sigma_r^2)$ and $\bar{\theta}_0 \sim N(\theta, \sigma_\theta^2)$, respectively, where r is the true initial target range and θ is the true initial bearing measurement. The initial target speed is initialized as $\bar{s} \sim N(s, \sigma_s^2)$, where s is the true initial target speed. Let $\sigma_{<.>}$ denotes the standard deviation of the parameter. Assuming that the target is moving towards the observer, the initial course estimate can be obtained as $\bar{c} = \bar{\theta}_0 + \pi$. The initial state vector and the initial covariance are given by

$$\hat{x}_0 = \begin{bmatrix} \hat{\zeta} \\ \hat{\eta} \\ \hat{\dot{\zeta}} \\ \hat{\dot{\eta}} \end{bmatrix} = \begin{bmatrix} \bar{r}\cos(\bar{\theta}_0) \\ \bar{r}\sin(\bar{\theta}_0) \\ \bar{s}\sin(\bar{c}) - \dot{\zeta}_0^o \\ \bar{s}\cos(\bar{c}) - \dot{\eta}_0^o \end{bmatrix}; P_0 = \begin{bmatrix} P_{\zeta\zeta} & P_{\zeta\eta} & 0 & 0 \\ P_{\eta\zeta} & P_{\eta\eta} & 0 & 0 \\ 0 & 0 & P_{\dot{\zeta}\dot{\zeta}} & P_{\dot{\zeta}\dot{\eta}} \\ 0 & 0 & P_{\dot{\eta}\dot{\zeta}} & P_{\dot{\eta}\dot{\eta}} \end{bmatrix} \tag{47}$$

where

$$P_{\zeta\zeta} = \bar{r}^2\sigma_\theta^2\cos^2(\bar{\theta}_0) + \sigma_r^2\sin^2(\bar{\theta}_0); P_{\eta\eta} = \bar{r}^2\sigma_\theta^2\sin^2(\bar{\theta}_0) + \sigma_r^2\cos^2(\bar{\theta}_0) \tag{48}$$

$$P_{\zeta\eta} = P_{\eta\zeta} = (\sigma_r^2 - \bar{r}^2\sigma_\theta^2)\sin(\bar{\theta}_0)\cos(\bar{\theta}_0); P_{\dot{\zeta}\dot{\zeta}} = \bar{s}^2\sigma_c^2\cos^2(\bar{c}) + \sigma_s^2\sin^2(\bar{c}) \tag{49}$$

$$P_{\dot{\eta}\dot{\eta}} = \bar{s}^2\sigma_c^2\sin^2(\bar{c}) + \sigma_s^2\cos^2(\bar{c}); P_{\dot{\zeta}\dot{\eta}} = P_{\dot{\eta}\dot{\zeta}} = (\sigma_s^2 - \bar{s}^2\sigma_c^2)\sin(\bar{c})\cos(\bar{c}) \tag{50}$$

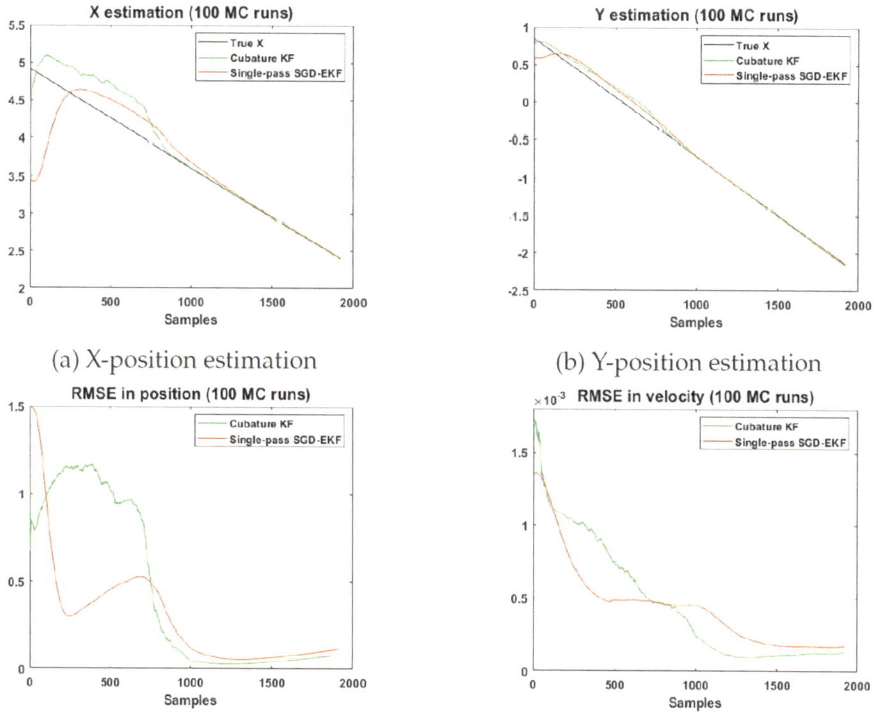

(a) X-position estimation

(b) Y-position estimation

(c) RMSE in position

(d) RMSE in velocity

Figure 4.
Comparison of estimation algorithms for the bearings-only tracking problem (100 MC runs).

where r and s are 5 km and 4 *knots*, respectively, and the target course is $-140°$. Here, σ_r is 2 *km*, σ_θ is 1.5 °, σ_s is 2 *knots* and $\sigma_c = \pi/\sqrt{12}$ for this problem.

Figure 4 shows a comparison of algorithms for the bearings-only tracking problem. The cubature Kalman filter (CKF) uses a third-degree spherical-radial cubature rule that provides the set of cubature points scaling linearly with the state-vector dimension [29]. The cubature Kalman filter and our single-pass SGD extended KF (EKF) method can track the target well, but our proposed method shows better computational efficiency compared to CKF by a factor of 2.5 (not shown). Root mean square error (RMSE) in position and velocity over 100 Monte Carlo runs are shown in **Figure 4c** and **Figure 4d**. During the whole maneuver, the RMSE of the proposed single-pass SGD-EKF algorithm was slightly lower than that with the CKF method.

4. Conclusions

In this chapter, we derived a single-pass sequential mini-batch SGD algorithm for estimating the noise covariances in an adaptive Kalman filter. We demonstrated the utility of the method using diverse examples involving a detectable (but not completely observable) system, a non-stationary system, and a nonlinear bearings-only tracking problem. The evaluation showed that the proposed method has acceptable state estimation root mean square error (RMSE) and exhibits filter consistency as measured by the normalized innovation squared (NIS) criterion.

Acknowledgements

This work was supported in part by the U.S. Office of Naval Research (ONR), in part by the U.S. Naval Research Laboratory (NRL) under Grant N00014-18-1-1238, N00014-21-1-2187 and Grant N00173-16-1-G905.

Abbreviations

CKF	Cubature Kalman filter
EKF	Extended Kalman filter
INS	Inertial navigation system
KF	Kalman filter
MMSE	Minimum mean square error
NIS	Normalized innovation squared
RMSE	Root mean square error
RMSProp	Root mean square propagation
SGD	Stochastic gradient descent

Author details

Hee-Seung Kim[1*], Lingyi Zhang[2], Adam Bienkowski[1], Krishna R. Pattipati[1], David Sidoti[3], Yaakov Bar-Shalom[1] and David L. Kleinman[1]

1 Department of Electrical and Computer Engineering, University of Connecticut, Storrs, CT, USA

2 HRL Laboratories, Intelligent System Laboratory, Malibu, CA, USA

3 U.S. Naval Research Laboratory, Marine Meteorology Division, Monterey, CA, USA

*Address all correspondence to: hee-seung.kim@uconn.edu

References

[1] Auger F, Hilairet M, Guerrero JM, Monmasson E, Orlowska-Kowalska T, Katsura S. Industrial applications of the Kalman filter: A review. IEEE Transactions on Industrial Electronics. 2013;**60**(12):5458-5471

[2] Bar-Shalom Y, Li XR, Kirubarajan T. Estimation with applications to tracking and navigation: theory algorithms and software. New York, NY: John Wiley & Sons; 2001

[3] Kalman RE. A new approach to linear filtering and prediction problems. Journal of Basic Engineering. 1960;**82**(1): 35-45

[4] Zhang L, Sidoti D, Bienkowski A, Pattipati KR, Bar-Shalom Y, Kleinman DL. On the identification of noise covariances and adaptive Kalman Filtering: A new look at a 50 year-old problem. IEEE Access. 2020;**8**: 59362-59388

[5] Kim HS, Zhang L, Bienkowski A, Pattipati KR. Multi-pass sequential mini-batch stochastic gradient descent algorithms for noise covariance estimation in adaptive Kalman filtering. IEEE Access. 2021;**9**:99220-99234

[6] Kim HS, Bienkowski A, Pattipati KR. A single-pass noise covariance estimation algorithm in multiple-model adaptive Kalman filtering for non-stationary systems. TechRxiv. 2022. DOI: 10.36227/techrxiv.14761005.v2

[7] Odelson BJ, Rajamani MR, Rawlings JB. A new autocovariance least-squares method for estimating noise covariances. Automatica. 2006;**42**(2): 303-308

[8] Mehra R. On the identification of variances and adaptive Kalman filtering. IEEE Transactions on automatic control. 1970;**15**(2):175-184

[9] Belanger PR. Estimation of noise covariance matrices for a linear time-varying stochastic process. Automatica. 1974;**10**(3):267-275

[10] Sarkka S, Nummenmaa A. Recursive noise adaptive Kalman filtering by variational Bayesian approximations. IEEE Transactions on Automatic Control. 2009;**54**(3):596-600

[11] Killick R, Fearnhead P, Eckley IA. Optimal detection of changepoints with a linear computational cost. Journal of the American Statistical Association. 2012;**107**(500):1590-1598

[12] Neethling C, Young P. Comments on "Identification of optimum filter steady-state gain for systems with unknown noise covariances". IEEE Transactions on Automatic Control. 1974;**19**(5):623-625

[13] Tajima K. Estimation of steady-state Kalman filter gain. IEEE Transactions on Automatic Control. 1978;**23**(5):944-945

[14] Bertsekas DP. Athena scientific optimization and computation series. In: Nonlinear Programming. Belmont, MA: Athena Scientific; 2016. Available from: http://www.athenasc.com/nonlinbook.html

[15] Golub GH, Van Loan CF. Matrix computations. Vol. 3. Baltimore, MD: JHU press; 2013

[16] Arnold WF, Laub AJ. Generalized eigenproblem algorithms and software for algebraic Riccati equations. Proceedings of the IEEE. 1984;**72**(12):1746-1754

[17] Battiti R. Accelerated backpropagation learning: Two

optimization methods. Complex Systems. 1989;**3**(4):331-342

[18] Boyd S, Xiao L, Mutapcic A. Subgradient methods. lecture notes of EE392o, Stanford University, Autumn Quarter. 2003;**2004**:2004-2005. Available from: www.stanford.edu/class/ee392o/subgrad method.pdf

[19] Tieleman T, Hinton G. Lecture 6.5-rmsprop: Divide the gradient by a running average of its recent magnitude. COURSERA: Neural networks for machine learning. 2012; **4**(2):26-31

[20] Kingma DP, Ba J. Adam: A method for stochastic optimization. arXiv. 2014. DOI: 10.48550/arXiv.1412.6980

[21] Zeiler MD. Adadelta: an adaptive learning rate method. arXiv. 2012. DOI: 10.48550/arXiv.1212.5701

[22] Kim HS, Zhang L, Bienkowski A, Pattipati KR. A single-pass noise covariance estimation algorithm in adaptive Kalman filtering for non-stationary systems. In: 2021 IEEE 24th International Conference on Information Fusion (FUSION). IEEE; 2021. pp. 556-563. DOI: 10.23919/FUSION49465.2021.9626861

[23] Naets F, Cuadrado J, Desmet W. Stable force identification in structural dynamics using Kalman filtering and dummy-measurements. Mechanical Systems and Signal Processing. 2015;**50**: 235-248

[24] Simon D. Optimal state estimation: Kalman, H infinity, and nonlinear approaches. Hoboken, NJ: John Wiley & Sons; 2006

[25] Woodman OJ. An introduction to inertial navigation. Technical Report, Computer Laboratory. Cambridge, UK: University of Cambridge; 2007 UCAM-CL-TR-696. DOI: 10.48456/tr-696

[26] Kuritsky MM, Goldstein MS. Inertial navigation. Proceedings of the IEEE. 1983;**71**:1156-1176

[27] Heller WG. Free-Inertial and Damped-Inertial Navigation Mechanization and Error Equations. Reading, MA: Analytic Sciences Corp; 1975

[28] King A. Inertial navigation-forty years of evolution. GEC review. 1998; **13**(3):140-149

[29] Arasaratnam I, Haykin S. Cubature Kalman Filters. IEEE Transactions on Automatic Control. 2009;**54**(6): 1254-1269

[30] Farina A. Target tracking with bearings – Only measurements. Signal Processing. 1999;**78**(1):61-78

[31] Aidala VJ. Kalman Filter Behavior in Bearings-Only Tracking Applications. IEEE Transactions on Aerospace and Electronic Systems. 1979;**AES-15**(1):29-39

[32] Leong PH, Arulampalam S, Lamahewa TA, Abhayapala TD. A Gaussian-Sum Based Cubature Kalman Filter for Bearings-Only Tracking. IEEE Transactions on Aerospace and Electronic Systems. 2013;**49**(2): 1161-1176

[33] Ristic B, Arulampalam S, Gordon N. Beyond the Kalman filter: Particle filters for tracking applications. Norwood, MA: Artech house; 2003

[34] Kumar K, Bhaumik S, Arulampalam S. Tracking an Underwater Object with Unknown Sensor Noise Covariance Using Orthogonal Polynomial Filters. Sensors. 2022;**22**(13):4870